ALASKA'S

WILDERNESS MEDICINES

Healthful Plants
of the
Far North

ALASKA'S

WILDERNESS MEDICINES

Healthful Plants of the Far North

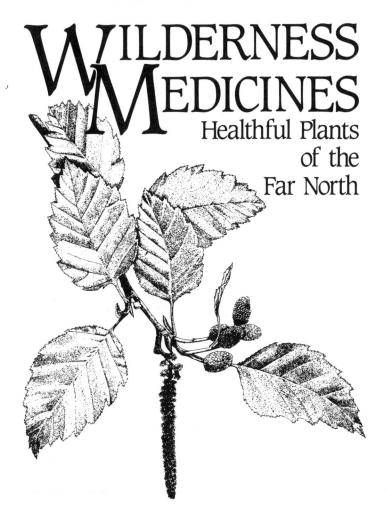

Eleanor G. Viereck

Illustrated by Dominique Collet

Alaska Northwest Books™

Anchorage • Seattle • Portland

First printing 1987
Fourth printing 1994

Library of Congress Cataloging-in-Publication Data
Viereck, Eleanor.
    Alaska's wilderness medicines.
    Bibliography: p. 97
    Includes index.
    1. Medicinal plants—Alaska. 2. Materia medica, Vegetable—Alaska.
I. Title. [DNLM: 1. Plants, Medicinal—Alaska. QV 770 AA5 V6a]
QK99.U6V54   1986   581.6'34.'09798   87-1427
ISBN 0-88240-322-2

Cover design by Robert Chrestensen
Book design by Martine Richards

Alaska Northwest Books™

An imprint of Graphic Arts Center Publishing Company
Editorial office: 2208 NW Market Street, Suite 300, Seattle, WA 98107
Catalog and order dept.: P.O. Box 10306, Portland, OR 97210
    800-452-3032

Printed on acid-free recycled paper in the United States of America

Contents

Acknowledgments

I wish to acknowledge the assistance and motivation provided by Patsy Turner in obtaining funds to prepare the manuscript and in typing the first draft. I also wish to thank Dr. Helen Bierne, Commissioner of the Department of Health and Social Services, State of Alaska, and Dr. Nils Annerud, Coordinator of the Preventive/Holistic Health Project. A previous edition of this book was made possible by a grant from the state of Alaska.

Grateful acknowledgment is made to Dr. Dave Murray, Director of the Herbarium, University of Alaska Fairbanks Museum, for allowing me the use of his specimens in writing descriptions of plants. Without the botanical authority and personal support and good humor of my husband, Leslie A. Viereck, this book would never have been written. And the editing contribution of Maureen A. Zimmerman of Alaska Northwest Books™ has been thorough and responsive; she showed me many ways to improve this book and I thank her very much.

About This Book

The purpose of this book is to acquaint people with Alaskan wild plants, native and introduced, which can be used to promote health and healing, for first-aid emergency care, or to maintain wellness.

I hope the book will be useful to persons in cities, on farms, and in the wilderness, whether they are in Alaska for recreation, hunting, fishing, or work. Others, inadvertently stranded as a result of accident or disaster, may find themselves in need of help from healing plants.

More than fifty plant species are described, with information on habitat and distribution and general information on how each one can be used as medicine.

There are some additional notes on history and folklore, poisonous species that might be confused with useful ones, and my own experiences with the plants. Information about the constituents of the plants is also presented.

This book is the first of its kind, as far as I know, integrating and combining Alaskan ethnobotanical lore with European and American herbal traditions to serve as a natural history guide to medicinal plants of Alaska and their uses.

It is far from complete in two important regards: First, it does not cover all of the medicinal plants of Alaska. There are lichens, seaweeds, and other herbaceous plants that may be very useful, but unfortunately I have not yet experimented with them and studied the uses of all of the species that may be of interest. And, there may well be plant cures not yet discovered. Second, I have deliberately omitted some plants because I consider them rare and endangered species and do not want to encourage their decimation by well-meaning gatherers.

And lastly, but most importantly, I wish to stress that this small natural history of a few of the Alaskan medicinal plants is not intended to serve the purpose of a self-care manual of medicine. I do not want any reader of this book to neglect a serious illness or injury that should have professional medical attention.

Collecting Wild Plants

The technique for gathering useful medicinal material from wild plants varies according to the season, the type of plant, and the part of the plant you intend to gather.

To collect flowers, fruits, buds, twigs, leaves, and some shallow roots, all you need is your hands. For deeper roots a shovel or spade is the best tool. For many plants a knife is required to cut the fibrous stems, and for the inner bark of shrubs and trees a large knife or small axe is usually the best type of tool to use.

Since lead often appears in the vegetation beside well-traveled highways (Pfeiffer), it is wise to avoid the convenience of roadside collecting and be prepared to walk off the beaten path in search of wild plants, to reduce the hazard of lead poisoning.

Sensitivity to the environment of our spaceship the planet earth is, alas!, not something I can take for granted. It has been brought to my attention that merely the appearance of this book on medicinal plants may result in over-utilization of rare species, degradation of sensitive ecosystems, and insults and abuses to local Native people and their relationship with their traditional plant resources. This consideration was serious enough that I have hesitated to contribute to the decline of the wild environment by encouraging great hordes of people to go stomping and tramping over bogs and tundra to dig up roots and leave a wake of devastation behind.

One way to reassure ourselves that our planet will remain habitable is to accept the hypothesis, recently termed the Gaia hypothesis but similar to the intuitive visions of sages and mystics in many times and places, that the earth is one organism. If this is at all true, then we are all parts of one entity and whatever we do to harm one part of it will ultimately harm our own selves. There is no conflict between the best interest of the whole and the best interests of one individual self if that self realizes the relationship between itself and the environment.

Sometimes it is a matter of heedless carelessness that leads people to destroy the habitat by digging and collecting more than they need. Sometimes the selfishness and thoughtlessness apparent in people who dig up the last specimen of the pasque flower to show it to their biology class is a sort of ignorance masquerading as pride, or hubris.

I know in my own life there is a steady tendency toward more careful consideration before I pick one wild plant. There are more and more square miles of space in which I do not feel comfortable harvesting certain rare and endangered species. And so my own subjective judgement and ethical considerations toward the rights of the plants and the rights of other animals and other humans to use the plants are subject to continual re-examination to decide what constitutes use and what constitutes abuse.

The American Indians would offer a prayer of thanks before sacrificing a medicinal plant. Rolling Thunder would even offer a sacrifice and conduct a ritual explaining to the plant why he needed it.

Surely it makes sense not to take all of the yellow dock or wormwood, or even yarrow or dandelion. It is true that many of the medicinal plants are "weeds" and are removed as pests anyway. Or maybe we are the parasites, in which case we might be wise to recall the stratagem of the successful parasite — to take care not to kill the host.

From Leaf to Tea

Select leaves that are not decayed or discolored and gather them when they are dry. Take care not to disturb the area, avoid leaving trash and litter behind, and do not take all of the plant; harvest only a portion of what there is. Take only what you need. Attention to these considerations has an effect on your own state of mind toward the healing virtues of the plant.

Drying and Storing Leaves

Lay the leaves on a rack or screen or hang them in small bunches from pegs or strings in an airy place that is as warm and dry as possible, but not in direct sunlight. If your climate is hopelessly damp and humid or the plants too succulent, you may need to use a warm oven to hasten drying.

After the leaves are brittle they should be stored in airtight glass or tin containers in a dark, dry place that is cool. Don't crumble the leaves into smaller pieces than necessary.

Making Tea

If you wish to extract more of the material from dried leaves, and you have a good fine paper or cloth filter, you may wish to use a mortar and pestle to grind the leaves into powder just before you make tea.

Otherwise, it is usually sufficient to break up the leaves into small pieces with your fingers and put in a preheated glass container. Pour freshly boiled water over the leaves, cover the pot or cup with a lid to keep the temperature high and to prevent aromatic materials from escaping with the steam, and steep for three to five minutes.

While the tea is steeping, and while you are sipping it slowly, it is highly recommended to be quiet — not reading, working, or bustling around. Most tonic teas are best taken on an empty stomach rather than with meals. And if there is a specific reason for using medicinal tea, meditating upon the healing you desire to take place while you sip the tea may promote the good effect.

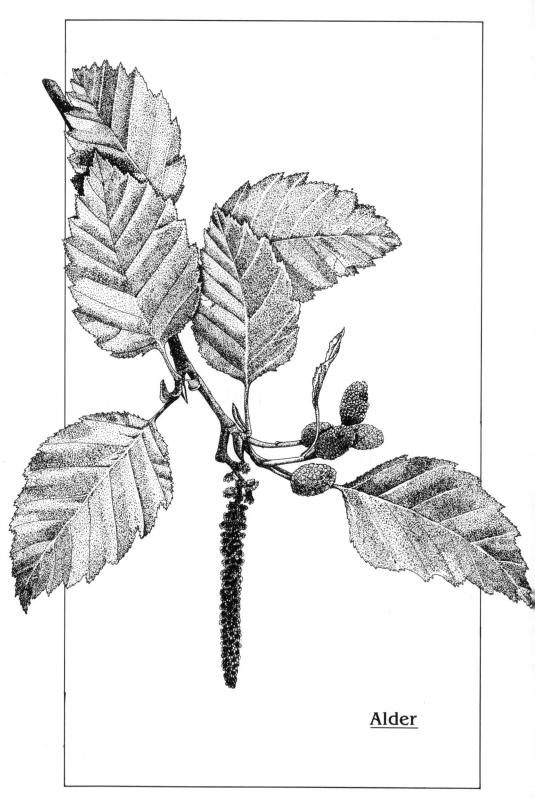

<u>Alder</u>

4

Alder

Alnus species
(Betulaceae)

Description:

The alders, of which Alaska has three species *(Alnus crispa, A. rubra,* and *A. tenuifolia)*, are large shrubs sometimes reaching tree size. (Viereck) Their stems are not erect and rigid, but tend to bend under the weight of the winter snow cover — an alder thicket becomes a maze of stems at various angles. This growth form is distinctive, whether the shrub grows alone or in a dense zone or band at treeline in the mountains.

Old fruits, conelike and dark brown, persist — they hang in clusters throughout the winter and into the next year. In summer appear leaves that are sharp-pointed at the apex, rounded or broadly wedge-shaped at the base; leaf edges are sharply and finely toothed, with long-pointed, nearly even or even teeth. Winter buds and summer leaves are slightly resinous, with a faint odor. The flowers are catkins and the roots, like those of legumes, often have root nodules — swellings containing berries that fix nitrogen from the air and enrich the soil.

Distribution:

Alders grow along roadsides, at treeline, in woods and meadows, and on nutrient-poor, newly exposed soil along rivers.

Constituents:

Low molecular-weight phenol, neurotoxin, and an insecticide are present, according to John Bryant. (personal communication)

Medicinal uses:

The inner bark of alder is Tanaina (Anchorage-area Native) medicine; the Natives boil the bark and drink the tea to get rid of gas in the stomach and to lower a high fever. (Kari) The astringent and powerfully bitter bark infusion is used as a gargle for sore throat, to induce circulation, to check diarrhea, and for eye drops.

Alder leaves are used to cure inflammation. Fresh leaves applied to bare feet are good for burning and aching; they are also used as a foot bath when brewed.

Bark of the red alder *(A. rubra)* of the coastal regions and western North America was used by Indians to relieve indigestion and as a tonic and alterative.

During the routine screening of southwestern Alaska plants for potential antitumor activity, the stem bark of *A. oregona* (same as *A. rubra)* showed significant antitumor activity. Lupeol and betulin were identified as the two constituents responsible for this activity. (Sheth et al.)

Other uses:

A fire made from green alder wood burns hot enough to use for welding. Alder twigs and buds make up an important part of the food of ptarmigan. In fall and winter the "seeds," or nutlets, are eaten by many songbirds.

5

Angelica

Angelica

Angelica species
(Umbelliferae)

Description:

The stems of these plants are 3 to 4½ feet (1 to 1.5 meters) tall, erect, hollow, and coarse, with many oil tubes. The compound leaves form groups of three leaflets; leaf stalks have inflated bases that sheath the stem. Leaflets are ovate, toothed and not hairy. The inflorescence, an umbel, has no involucre. The flowers are greenish white.

The ribbed fruits of *Angelica genuflexa* have wings and the upper leaves are reflexed. On the other Alaskan species of Angelica, *A. lucida,* the ribbed fruits are not winged; the upper leaves are not reflexed.

Distribution:

Angelica lucida grows in coastal Alaska and the Alaska Range, eastern Siberia, and Canada. *Angelica genuflexa* is less widely distributed, occurring in the coastal zone of southeastern and central Alaska and on the Aleutians. Other species of *Angelica* appear in China and Europe.

Constituents:

Information is available only for *A. archangelica,* a European species. A volatile oil, angelic acid, and resin are listed in *The Merck Index.* Taskinen and Nykanen describe a large number of constituents including g-phellandrene, borneol, pentadecanolide, 2-methyl butyric acid, and monoterpene hydrocarbons. The musklike odor of angelica root oil is attributed to the lactone of hydroxypentadecanoic acid.

Medicinal uses:

Tanainas use *A. lucida* root as a medicine on the outside of the body — for aches, pains, sores, cuts, blood poisoning, and any kind of infection. They first cut up the root and mash it, then boil the mashed root or soak it in hot water. Then they place it on the place needing treatment. (Kari) Older Eskimos would slice the roots into two parts, heat the halves and place them over the area of the body that hurt, even for deep pain. (Smith) One of the few Alaskan plant medicinal use citations in Hulten concerns *A. lucida:* The Siberian Eskimos inhaled the fumes of the roasted root as a seasickness remedy.

Angelica leaves were used as a poultice by the island and Iliamna Tanainas.

European *A. archangelica* virtues were praised in old writings. The plant name itself testifies to the great antiquity of a belief in its merits for protecting against contagion, for purifying the blood, and for curing every conceivable malady. It was held a sovereign remedy for poisoning, indigestion, general debility, agues, and all infectious maladies. (Grieve, Simmonite-Culpeper) Its therapeutic actions are carminative, diaphoretic, diuretic (Merck), expectorant, emenagogue (Grieve), stimulant, stomachic, tonic. An infusion from the root is used to relieve flatulence.

A sample of *A. sinensis* root slices purchased in China seemed to me to have an odor very similar to the Alaskan *A. lucida.* Another Chinese *Angelica* is described by Hong-Yen Hsu. This plant, *"Angelicae Dahuricae Radix,"* was first recorded as a "drug" in one of the oldest Chinese books, *The Classic on Mountains and Seas* (400 B.C.). Hsu says it can be used to treat the common cold, migraine, dizziness, neuralgia and excessive perspiration. Jackson and Teague say the Chinese herb *dong quai,* obtainable in powdered, sliced or whole chunk form in many herb stores, is *Angelica,* and is a female tonic similar to the tonic for men, ginseng. (Devil's club is known as Alaska ginseng, since it is in the same family.) A friend has reported that for the first time in her life, after drinking dong quai tea daily for about one month, her menstrual period has become regular and pain-free. *Angelica polymorpha* var. *sinensis* extracts have long been used in Chinese medicine to stimulate uterine contractions as a post-partum aid.

The juice of *Angelica* is used in both Eurasia and America to relieve pain in a decayed tooth, and it is one of the many herbs considered a tonic to improve well-being and mental harmony. (Lewis and Elvin-Lewis)

Other uses:

The ripe fruit of *A. archangelica* is used in teas throughout the world. Its root is used to flavor cigarette tobacco. Angelica root is grown commercially; its derivatives (the essential oil, extracts, distillates) find extensive use in flavor formulations for alcoholic beverages such as vermouths, bitters, benedictine and chartreuse-type liqueurs. (Taskinen and Nykanen)

Although angelica is included in the list of higher plants that provoke photodermatitis in man, it is one of the many herbs used for perfume and for scenting creams, salves, soaps, oils and shampoos.

I have had difficulty in getting angelica seeds purchased from an herb company to germinate, as they evidently require light (seeds usually require dark to germinate).

Warning:

There is a tendency for angelica to increase sugar in the urine, so it would be contraindicated in persons with diabetes.

Before attempting to gather angelica, you should be sure that you know how to distinguish it from the poisonous water hemlock, *Cicuta.* They both have white umbels and some of the species have similarly shaped leaves. The *Cicuta* root has large transverse chambers separated by shelves (see illustration below).

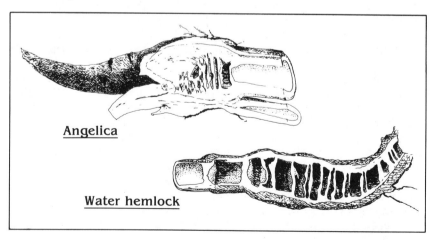

Angelica

Water hemlock

Birch

Betula papyrifera
(Betulaceae)

Description:

Alaska has three kinds of tree birch and two kinds of dwarf birch (only the tree birches have the uses described here). The tree birches hybridize wherever they meet, so they are considered three geographical varieties of a single transcontinental species.

The white, paperlike bark of these trees separates into layers. The bark remains smooth as a result of its persistent cork. Twigs growing above the reach of browsing moose have a smooth bark, but the lower ones are covered with white bumps called lenticels. The lenticels are believed to be associated with defense against browsing herbivores.

Each alternately arranged leaf is round with a sharp point. Leaf margins are sharply toothed with teeth of two sizes. The flowers form catkins with long, narrow male and short female flowers. Conelike fruits have many nutlets, seeds, and scales. Winged birch seeds often cover the snow with their tan crosses.

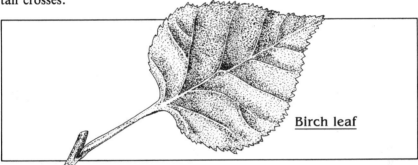

Birch leaf

Distribution:

If one includes the dwarf birch, *Betula nana,* the birch is distributed over all of Alaska except the very northern coast and the tip of the Aleutians.

Constituents:

The Merck Index cites betulin (betula camphor) 10% to 15% in the outer portion of the white bark. Leaves contain betuloresinic acid, essential oil, ether, betuloside, gaultherin, methyl salicylate, and ascorbic acid; in the bark of the sweet birch is salicylic acid.

Medicinal uses:

Margaret Lantis reports Natives using birch leaves to make a comforting tea. Birch leaf tea has been used as therapy for gout, rheumatism, and dropsy and also for dissolving kidney stones. Simmonite-Culpeper describes birch as a diuretic if a strong juice is made from the leaves. A decoction of leaves may be used as a mouthwash. According to Grieve the young shoots and leaves secrete a resinous substance that, combined with alkalis, forms a laxative.

Birch

A birch bark decoction can be used for bathing skin eruptions. The inner bark is astringent and bitter; it has been used to treat intermittent fevers.

Birch sap as medicine and spring tonic is bottled and sold in Russia. (L. Viereck, personal communication) Kari reports the Tanainas put fresh birch sap on boils and sores. The old way to obtain sap is to peel back the bark and scrape or suck the sap off the wood.

Culinary uses:

Birch sap can be boiled down to make a sweet syrup. My method of collecting birch sap is to use the equipment manufactured and sold to the maple syrup industry. The equipment consists of a spout that is inserted into a hole drilled into the tree, and a bucket that hangs from the spout to catch the sap as it drips out. I have made a ten-minute videotape documentary of this procedure, available from Teri Viereck, 1707 Red Fox Drive, Fairbanks, AK 99701. The price is $25.

If you want to drill a hole in a birch tree to collect sap, wait until the sap is flowing or the surface will scar over and reduce the flow. The sap may begin flowing any time from mid-April to mid-May. You can cut a small branch or drill a test hole to find out if the sap is flowing. Sap flow is the result of "root pressure" — that is, water is actively absorbed by the root system, but pressure builds because little water is lost from the tree as a whole. The sap flows the week or two before the leaves begin to open; once they are growing, the root pressure lowers. Near the end of the season the sap gets white, milky, and bitter due to yeasts.

Birch sap contains the sugars glucose and fructose, whereas maple sap contains mostly sucrose. Another difference is that birch sap is more dilute than maple — birch sap has only 1% syrup by weight.

To make a thick syrup, birch sap must be boiled down to decrease its volume by thirty or forty times, but maple sap (2.5% sugar) needs only twenty-five times concentration. It is very difficult to reach the syrup stage without burning the sap, so use a very very low fire, an asbestos pad, or steam (as in a double boiler) and watch it carefully. I have spent days of boiling and wound up with a bit of charcoal more than once. This year I let it go a bit too far over the cool end of my wood stove and obtained a delicious taffylike candy that tastes like molasses.

Some people use birch sap to make beer, wine or soft drinks.

Other uses:

A brown dye can be obtained from birch bark. The bark also yields oil of birch tar, which imparts durability to leather. The white epidermis of the bark can be separated and used as a substitute for oiled paper.

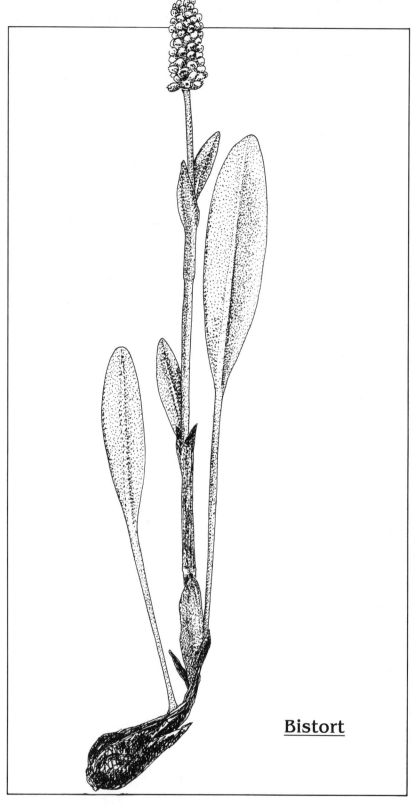

Bistort

Bistort

Polygonum bistorta
(Polygonaceae)

Description:

The stem of bistort forms a stalk 1½ to 2¾ feet (45 to 85 cm.) tall. The leaves, arising from the base of the stem, are elliptical or elongated with rounded or mostly cuneate bases and winged petioles. They are bluish green above with a purplish tint underneath, somewhat long and broad — resembling sourdock leaves.

The flowers, pink with dark anthers, are borne on a dense, cylindrical spike on the top of the stem.

Bistort root is a rhizome, thick and hard, usually contorted or S-shaped, bent upon itself (bistorted), 3 inches (8 cm.) long and ¾ to 1 inch (2 to 3 cm.) thick. Externally the rhizome appears blackish, but it is reddish brown within. It is depressed or channeled on the upper surface and marked with convex root scars on the under surface. Around the rhizome, a ring of small woody wedges encloses a pith the same thickness as the bark. Rootstalk and leaves spring from the numerous black thready rootlets.

Distribution:

Bistort is circumpolar to circumboreal; it is found in Eurasia and Alaska in alpine meadows and tundra.

Medicinal uses:

The rhizome is the portion used in medicine.

Bistort is one of the most powerful astringents in botanic medicine. It is useful and effective for cleansing canker and morbid matter from the mucous membranes of the alimentary canal and at the same time toning the entire tract. It influences the kidneys, firms up and tones the tissue, both internal and external, and is a versatile styptic agent for hemorrhages.

For sore and ulcerated mouth or gums, nasal problems, and running sores, use a decoction or infusion of bistort rhizome as a rinse or wash. Insect stings and snake bites should be treated externally with a strong decoction; stronger decoctions should be taken internally. For regulating or decreasing the menstrual flow, douching with a diluted decoction has been recommended. For the treatment of leucorrhea, douche with a strong decoction.

For cuts and wounds, apply and powdered root directly to the injured part.

It is noted in the Chinese book, *A Barefoot Doctor's Manual,* that bistort clears fevers, detoxifies, loosens congestion and reduces swelling.

Chamomile

Chamomile

Matricaria matricarioides
(Compositae)

Description:

Chamomile's head, or capitulum, of small yellowish green flowers forms a rounded cone. The bracts of scales around the base of the flowers have thin, transparent (scarious) membranes along their edges. There are no ray flowers or "petals" such as one finds in the aster, daisy, and another "chamomile," *Anthemis*. The leaves, finely dissected in featherlike lobes, may be more or less pointed at the ends. The round and furrowed stem measures 3 inches to 1 foot (8 to 30 cm.) tall. This particular species of chamomile is called pineapple weed because it smells like pineapple.

Distribution:

Matricaria matricarioides is a cosmopolitan weed. (Hulten) It seems to share the property of the other chamomiles *(Anthemis nobilis* and *Matricaria chamomilla)* of growing on seldom-used paths and roads.

Constituents:

Three recent papers have been published on the substances contained in German chamomile, *M. chamomilla*. Saleh found that red light increases the production of flower heads, while green light has a better effect on the quality of heads and percent of essential oil and chamazulene. Felklova and Jasicova published an article on the substances contained in *M. chamomilla* in 1975 in Czechoslovakia. And Padula, Rondina, and Coussio determined essential oil, total azulenes, and chamazulene in *M. chamomilla* cultivated in Argentina.

Medicinal uses:

The Tanainas boil the whole above-ground portion of the plant in water, strain the tea, and give it to a new mother and her baby to drink. They say it cleans them out and helps the mother's milk start. The Kenai Tanainas give the tea to anyone who needs a laxative and use it as a wash for the eyes and skin. (Kari) The Aleuts and Russians used the tea as a cure-all. (Smith)

I have found that tea made from the whole plant or the flowers with the leaves attached is very bitter indeed. A tea made from the flowers alone, however, tastes fragrantly sweet and quite similar to familiar chamomile tea.

Anthemis, the commercial chamomile, has been used to treat colds, fevers, painful and congested menstruation, bilious headache, indigestion, colic, spasms, coughs, bronchitis, pulmonary catarrh, acute dyspepsia, hysteria, nervousness, torpid liver, delirium tremens, rheumatism, ulcers, stomach weakness, ague, dropsy and jaundice. It has also been used when there are kidney, spleen, or bladder problems. It has been taken to expel worms, used as a wash for eyes and open sores, and in a poultice for swellings and toothaches.

Christopher describes *Anthemis* as diaphoretic, stomachic, tonic, antispasmodic, stimulant, sedative, emmenagogue, anthelmintic, anodyne, bitter aromatic, emetic (warm large dose), and cathartic (large dose).

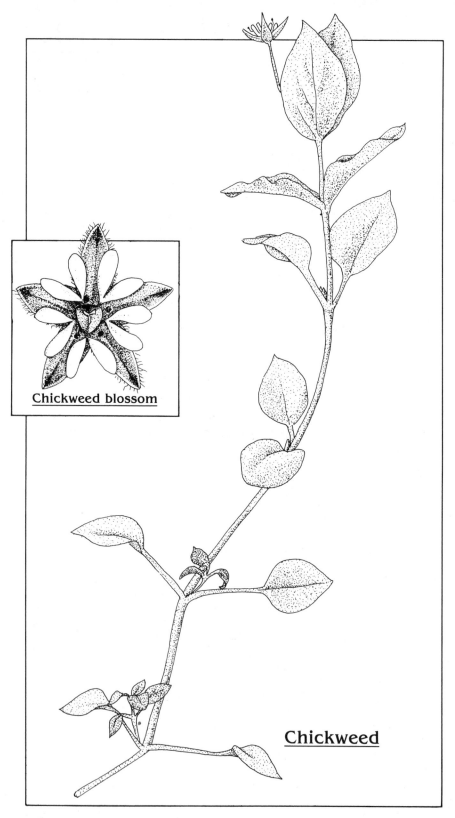

Chickweed blossom

Chickweed

Chickweed

Stellaria media
(Caryophyllaceae)

Description:

Chickweed stems, usually less than 1 foot (30 cm.) tall, are so weak that the whole plant leans on other plants or adjacent firm supporting surfaces, often forming a tangled mass. Wherever the stem touches the ground the nodes give rise to roots and new stems. The stem has a row of hairs that change sides at the nodes.

The opposite leaves are ovate to oblong in shape. The higher leaves are largely sessile, while those lower down on the stem have hairy petioles. The flowers, about ⅛ inch (2 to 3 mm.) in diameter, consist of five petals, but each petal is so deeply cleft that the blossoms look ten-pointed. The white petals are shorter than the green sepals.

Distribution:

Occurs worldwide. Very common around dwellings.

Constituents:

The main components are potash salts and vitamin C.

Medicinal uses:

The action of chickweed is demulcent, refrigerant, and antiscorbutic. Chickweed, called the magic healer by Gibbons, is my favorite herb to "draw the poison out" of infections, inflammations, boils, or abcesses. Any form of internal inflammation is soothed and healed by application of chickweed as an external poultice.

Fresh chickweed leaves are so soft and tender that it is possible to macerate them with your thumbnail into a green pulp. Apply this pulp fresh every hour or so and hold it in place with a bandage. Some herbalists prefer to cover the fresh herb with boiling water, cool it a bit, and then apply.

Chickweed is also used for burns, skin diseases, sore eyes, and wounds. (Leek) A salve of chickweed powder may be applied in addition to bathing with chickweed tea and drinking the tea.

One old astrological herbalist says of chickweed, "It is a fine, soft, pleasing herb, under the dominion of the Moon. In a word it comforteth, digesteth, defendeth, and supporateth very notably."

A vaginal bolus formula from Christopher that I have modified and used includes chickweed along with squaw vine herb, slippery elm bark, comfrey root, yellow dock root, golden seal root, mullein, and marshmallow root, with cocoa butter to hold them all together. Wrap with gauze to form a tampon and replace three times a day.

Culinary use:

Chickweed is fine to eat raw or cooked.

17

Cleavers

Cleavers

Galium boreale
(Rubiaceae)

Description:

Cleavers, also called lady's bedstraw, baby's breath, or goosegrass, is distinguished by the arrangement of the leaves on the square, 2-foot (60 cm.) tall, erect stem. Leaves are in a whorl, which means that four or five of the narrow, smooth-margined, sharp-pointed leaves are attached to the stem at one spot. A few centimeters farther along the stem occurs another whorl of leaves. The leaves themselves have three veins and no stem; they are sessile. The numerous flowers form in loose panicles; each flower has four white petals. The fruit has hairs. Cleavers belongs to the Madder family, which includes coffee, quinine, ipecac, gardenia and woodruff.

Distribution:

Galium is distributed in temperate regions around the world. It is quite common as a garden weed and in the forests.

Constituents:

The plant contains starch, chlorophyll, and three distinct acids: a kind of tannic acid named galitannic acid, citric acid and rubichloric acid. It is probably the acid content that is responsible for the plant's property of curdling milk. The association of this plant with milk is reflected in its name, *Galium,* from the Greek word *gala,* milk, and the common name of cheese rennet.

Medicinal uses:

Cleavers is a soothing, relaxing diuretic that influences the kidneys and bladder and acts mildly on the bowels. Cleavers tea, a mild laxative, is also given for diarrhea — probably due to the astringent effect of the tannic acid.

One of cleavers' medicinal values, to promote the loss of weight (which seems to take about six weeks), is probably due to the ability of its acids to speed up the metabolism of stored fat.

The plant works best if you gather your own herbs rather than buy them in the herb store. Be forewarned, however — continued use of cleavers tea has caused irritation to the mouth and tongue. To prevent this sort of irritation, a demulcent smoothing ingredient should be added to the formula. Slippery elm and marshmallow root have been used for this purpose.

Cleavers should not be given where there is a tendency toward diabetes.

Since cleavers is a refrigerant herb (Christopher) it would not be indicated for a type of person who is always cold. But, it is good to apply to the face with cotton to cool a sunburn. Euell Gibbons recommends the tea as a lotion on the face to remove freckles and to make the complexion clear.

Simmonite-Culpeper recommend a cleavers decoction to stop inward bleeding and to heal all inward wounds generally.

The Tanaina use cleavers the same way they use wormwood *(Artemisia tilesii).* Lime villagers say it is especially good as a hot pack for aches and pains. (Kari)

19

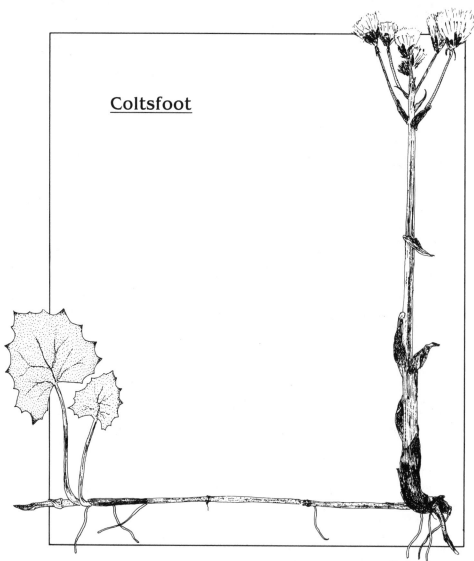

Coltsfoot

Coltsfoot

Petasites species
(Compositae)

Description:

Since coltsfoot flowers appear before the leaves they must be collected separately. A single stalk bears one or several flowers, purple and covered by numerous long white hairs. After the flowers have died down, the leaves sprout and continue to grow in size throughout the summer until they are nearly a foot (30 cm.) across. They are triangular in basic outline, but their edges are cut into lobes and teeth. The top surface shines while underneath the color is silvery white, the texture woolly. Coltsfoot's small white root is shallow.

Petasites frigidus, also called *Tussilago frigida,* has shallowly lobed, kidney-shaped basal leaves. It is described as being from Lapland, Switzerland and Siberia. According to Hulten its root is eaten, roasted, by the Siberian Eskimos. *Petasites hyperboreus,* which hybridizes with *P. frigidus,* has more triangular, deeply lobed leaves. *Petasites sagittatus,* a North American species, displays triangular, arrow-shaped leaves. These are the common Alaskan species.

Distribution:

Coltsfoot is found in bogs, meadows and wet places.

Constituents:

According to Spoerke, *Tussilago farfara* (a related species), called coughwort by European herbalists, contains a bitter glycoside, tannin, caoutchouc, a saponin, a volatile oil, a resin and pectin. Grieve says this species contains mucilage, a little tannin, and a trace of a bitter amorphous glucoside. The flowers also contain a phytosterol and a dihydride alcohol.

Medicinal uses:

The dried leaves or flower shoots have been used since ancient times against persistent cough. One common name for coltsfoot is coughwort, and tussilago means cough dispeller. For bronchitis, the dried leaves have been smoked since the time of Pliny, who used a hollow reed for the purpose. I have tried smoking coltsfoot leaves either by inhaling the smoke as the leaves came into contact with charcoal, or actually smoking the crushed leaves in a pipe alone or mixed with other herbs. The cooling effect seems to me somewhat like the menthol in a cigarette and actually does seem to have a calming and enlarging effect on the bronchial passages, although it is hard to recommend smoking as a therapeutic practice when there is so much lung illness attributed to smoking.

Tanainas soak coltsfoot root in hot water and drink the tea for tuberculosis, chest troubles, sore throat, and stomach ulcers. Old-timers also drank the tea "to make the blood soft" and chewed the root for tuberculosis. (Kari) The raw or boiled root of *P. speriosus* (a non-Alaskan relative) is used as cough medicine by the Quileute of Washington state. (Gunther) "They also mash the root and soak it as a wash for swellings and sore eyes. The Skagit warm the leaves and lay them on parts afflicted with rheumatism." (Gunther) The root decoction was used against asthma or rheumatism. (Tobe)

The leaf or flower infusion or tincture is used against diarrhea, probably due to its astringent action. Crushed coltsfoot leaves or a decoction can be applied externally for insect bites, inflammations, general swellings, burns, erysipelas, leg ulcers, and phlebitis. (Lust)

Besides being astringent, the infusion of leaves or flowers is demulcent, emollient, expectorant, pectoral, diaphoretic, and tonic. (Kloss) It is excellent in a cough syrup with other herbs and is also good as a fomentation. Or, just moisten a cloth with coltsfoot tea and apply as a poultice to relieve the chest of phlegm.

Warning:

Heinerman cautions that large amounts and strong doses of coltsfoot tea may produce an abortion in a pregnant woman.

21

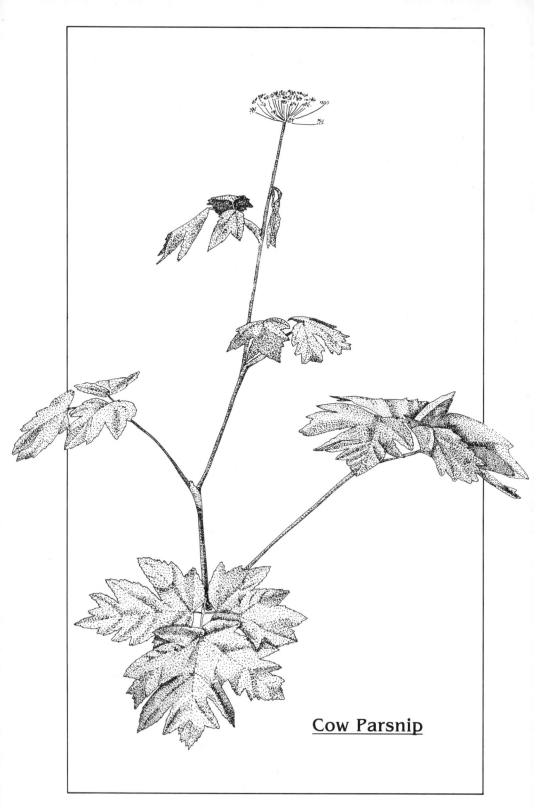

Cow Parsnip

Cow Parsnip or Wild Celery

Heracleum lanatum
(Umbelliferae)

Description:

The cow parsnip, also called wild celery, is a large plant, up to 10 feet (3 meters) tall, with a stout grooved stem and large compound leaves divided into three leaflets. Each leaflet is ½ to 1½ feet (15 to 50 cm.) wide, somewhat maplelike in shape with coarse, irregularly cut margins. The stem and leaf stalks are densely hairy; they clasp the main stem at the base. Flowers form large, broad umbels, often 14 inches (36 cm.) or more across and flat-topped. Individual flowers are very small and white.

Distribution:

This species occurs around the northern Pacific Ocean up into interior Alaska and across Canada. (Hulten) It is common along roadsides in the Alaska Range.

Constituents:

The plant contains an unidentified volatile oil. (Spoerke) To me its odor is powerful and distinctly medicinal.

Two species of *Heracleum* from the Himalayas contain coumarins that can be used to produce xanthotoxin, a dermal photosensitizing agent. (Kumar, Banerjee & Handa)

Medicinal uses:

The fruit, green parts, and root of cow parsnip are all pulverized to make poultices and tonics. For arthritis, grease the root, steam it, and split it. Leave it on the affected spot overnight. (According to de Laguna, cow parsnip root is used by the brown bear when wounded.)

Warren Smith reports that cow parsnip is much used in tonics for colds and sore throats. The root is Tanaina medicine for colds, sore throat, mouth sores, and tuberculosis. Chew the root raw or boil it in water and drink the tea.

Hall says to eat cow parsnip to calm the nerves.

Densmore cites the leaves and stem as a rubefacient, the roots as carminative and stimulant in the tradition of the Chippewa Indians.

Culinary uses:

Be sure to peel cow parsnip stem before you eat it, as the hairs are irritating to the mouth. And be sure to distinguish this plant from the poisonous water hemlock that it resembles (see illustration on page 8). Young leaf stalks and stems can be eaten raw, although cooking improves the flavor.

Crowberry

Crowberry

Empetrum nigrum
(Empetraceae)

Description:

This low, creeping or spreading evergreen, heatherlike shrub forms dense mats with horizontal, much-branched stems. Crowberry leaves are usually needlelike, crowded, four in a whorl; sometimes they are alternate, about ¼ to ½ inch (3 to 6 mm.) long, shiny, yellow-green with a groove on the lower surface formed by curved margins. The upward-curving twigs are 2 to 6 inches (5 to 15 cm.) long, brown, very slender, and finely hairy becoming shreddy.

The purplish flowers — single, inconspicuous, and stalkless — sprout at the base of the leaves. Each small, ¼ to ½ inch (3 to 6 mm.) flower is composed of three bracts, three sepals, three spreading petals, and three stamens much longer than the petals. Each pistil bears a six to nine-celled ovary and flat stigma with six to nine narrow lobes.

The fruits, round and berrylike, measure about ⅜ to ½ inch (5 to 10 mm.) or more in diameter. Shiny, dark blue-black or purple, very juicy and sweet, they contain six to nine reddish-brown nutlets.

Distribution:

Empetrum nigrum is found throughout Alaska in forests and muskegs and on the tundra.

Medicinal uses:

The leaves and stems are Tanaina (Anchorage-area Native) medicine for diarrhea. Boil or soak crowberry leaves and stems in hot water and drink the tea. Some Inland Tanainas say the berry juice is good for kidney trouble. The root is medicine too and is used for sore eyes and to get rid of cataracts. Boil the roots and wash the eyes with the cooled juice. Bark from the stem is good for the eyes, too. (Kari)

Other uses:

Crowberry is used as a ground cover in Interior Alaska. Plants can be grown from cuttings.

Currant

Currant

Ribes species
(Grossulariaceae)

Description:

The genus *Ribes,* with seventy-five species in North America, includes the currants and the gooseberries. These shrubs have maplelike leaves. The berries are black to purple with gland-tipped hairs *(R. lacustre,* swamp gooseberry or swamp currant), glandular with a whitish to bluish bloom and fetid odor *(R. bracteosum,* the stink currant), black and bitter with resin dots *(R. hudsonianum,* the skunk currant), black with bluish bloom and gland-tipped hairs *(R. laxiflorum,* the trailing black currant), or translucent and red *(R. triste,* the American red currant).

Usually the flowers and berries are borne in drooping racemes. The small flowers have a tubular base with five sepals larger and more conspicuous than the five scalelike petals, and five stamens alternate with the petals.

Distribution:

The only places in Alaska that do not contain some type of currant are the extreme northern and western coasts and the Aleutians.

Medicinal uses:

According to Kari, the stem of *R. triste* is medicine for colds, flu, and tuberculosis. First scrape off the outer bark and throw it away. Then boil the inner bark and stem together. Drink the tea. Also, cooled tea is a warm wash for sore eyes. De Laguna reports similar use by the Mount Saint Elias people.

According to Kirk, the berries of *R. cereum* were used by the Hopi Indians to relieve stomachaches. And from the European herbalists, as noted in Simmonite-Culpeper, the jelly made with the juice of *R. vulgaris* is "cooling and grateful to the stomach" and cooling in fevers.

Washington state natives have many medicinal uses for *Ribes,* described by Gunther. Roots of the common gooseberry, boiled into a tea, are drunk by the Swinomish for sore throats, tuberculosis, and venereal disease. The bark is soaked to make an eyewash. The Cowlitz burn and pulverize the woody stem and rub the charcoal on sores on the neck. Bark of the swamp gooseberry or swamp currant, *R. lacustre,* is peeled off and boiled into a tea to drink during labor or to wash sore eyes. The Lummi make tea from the twigs for general aches. Bark, leaves, twigs, and roots of the trailing black currant, *R. laxiflorum,* were used to make medicinal tea for colds.

Note: The blister rust fungus that kills the white pine and other five-needle pines requires *Ribes* for part of its lifecycle. In eradication efforts many currant plants have been destroyed.

Dandelion
Taraxacum officinale
(Compositae)

Description:

This ubiquitous and cosmopolitan lawn and garden weed is also one of the most ancient remedies in the herbalist's repertoire, as the species name suggests. The common name, dandelion, stems from the French "dent de lion" or lion's tooth.

Its sunny yellow flowers that turn into white balls of airborne seeds are familiar to everyone I have encountered, although not everyone has dug the plant up to observe the vertical taproot furnished with numerous short, thickened rootlets. The leaves all attach at one point, the place where the stem and the taproot meet at the surface of the soil; they form a circle or basal rosette. Leaves are variously lobed and toothed and may be less than 2 inches to almost 2 feet (5 to 50 cm.) long, depending on habitat.

Dandelion

Method of collection:

Early June is the best time to harvest dandelion greens. I dig the whole plant from the garden with a shovel; it is easiest if the soil is soft from recent rain. Knocking the extra dirt from the roots, or picking away clods of dirt with fingers is easy enough, but if you want to use the greens and edible buds they are rather difficult to clean. Try to be careful not to get dirt on them. A strategy I discovered to minimize the spreading of dirt onto greens and buds was to collect small quantities at one time, process them, and then go back to dig for more. In a huge pile of uncleaned plants the dirt would inevitably be shaken from the roots onto the leaves.

The final cleaning of the roots works best with a dry brush, about 9 inches long, with big bristles. You do not want to use water, because wetting the roots would slow down the drying. Hitting the root with the tips of the stiff bristles is a good way to get it just as clean, and also removes the slender rootlets that protrude from the taproot. Slice the roots into ½-inch (1 centimeter) pieces. I have found it much easier to dry the dandelion roots if they are chopped; if I do not chop them into chunks when they are fresh it is almost impossible to do it later when they are hard and dry. The chopped roots dry in three days if they are spread out on a net, screen or drying rack.

Sun-drying does not damage roots as much as it damages leaves or stems. A warm, dark, well-ventilated place in a house or shed is all right. The warming oven of a woodstove is not too hot; in fact, slight roasting improves the sweet taste of dandelion roots. Some herb collectors are now using microwave ovens to dry their harvest.

The number of species is large, with many groups that are only recognized by specialists, but the one called *Taraxacum officinale* is the most widespread. Originally from Europe, it was introduced into South Africa, South America, North Zealand, Australia, India, and all of North America except the tundra barrens, although another species grows in the high Arctic tundra. Like most introduced weeds it is common in waste places, roadsides, and gardens.

Medicinal uses:

Dandelion has been considered a mild detergent, aperient, and diuretic.

The Latin name, *officinale,* indicates that dandelion was the official remedy for a number of ailments. It was an ingredient in many of the patent remedies of the snake-oil peddler days. To quote Millspaugh, "Taraxacum has been used in medicine from ancient times; it is one of those drugs, overrated, derogated, extirpated, and reinstated time and again by writers upon pharmacology, from Theophrastus to the present day."

Dandelion has been recommended in hepatic obstruction, and as a general liver and kidney stimulant. Since liver and kidney disorders manifest themselves in numerous ways, the dandelion root juice and extracts have been popular as a general tonic over the entire world. The plant's slightly bitter taste seems medicinal to many people.

Devil's Club

Devil's Club

Oplopanax horridum
(Araliaceae)

Description:

A prickly shrub with long, decumbent and often entangled branches or stolons, devil's club has dense spinose stems, petioles, and leaves. Each leaf is cordate at its base, deeply or shallowly five to seven-lobed. The lobes can be acute or with tails. The inflorescense, shorter than the leaves, carries umbels of green and white flowers. The fruits are inedible orange berries. The thick taproot contains a soft, pithy inner bark.

Distribution:

Devil's club is common as undergrowth in southeastern Alaska on moist well-drained soil, forming impenetrable thickets in coastal and flood-plain forests.

Medicinal uses:

The Kenai Tanainas boiled the stems and branches, then drank the resulting decoction for fever. The Upper Cook Inlet people boiled the inner bark of the underground portion of the plant and drank the tea for tuberculosis, stomach trouble, and colds as well as fever. The same inner bark is said to have been a treatment for swollen glands as well as boils, sores and other external infections. After it was baked slowly until it was very dry, it was rubbed between the hands until it was broken and soft. This pulp was then placed on the affected area to draw out infection. (Kari)

The Chugach Eskimos of Prince William Sound employed the ashes of *Oplopanax horridum* to treat burns. (Birket-Smith, cited in Kari)

De Laguna states that devil's club was "perhaps the most important medicinal and magical plant" of the Yakutat Tlingit. The shaman as well as the layman chewed the stem bark (scraped of its thorns) for its emetic and purgative effects as well as for a general cure-all. These people also drank devil's club infusions and used the bark as poultices and on hot rocks in the bath house.

The Haida people were still using devil's club for a variety of complaints in 1965. (Justice, cited in Kari) They would mix the dried inner bark with cedar or spruce pitch for a waterproof dressing for wounds. Other groups of people on the Northwest Coast have used the plant for medicine and to obtain supernatural power.

The modern medical world has been interested in devil's club since the 1930s, due to the discovery of the possible presence of an insulinlike substance in the plant. It does seem to be of value in maintenance of diabetes but its chemistry is still under investigation. The therapeutic action is hypoglycemic. (Lewis) (Kari and Smith, unpublished)

Fireweed

Fireweed

Epilobium angustifolium
(Onagraceae)

Description:

Fireweed raises a tall, simple, densely leafed stem from woody roots. Its leaves are alternate, lanceolate, acute and, underneath, pale and distinctly veined. The flowers, in long, terminal racemes opening from the base, display more or less red-colored sepals and large, clawed petals. Normally lilac-purple, the flowers are sometimes white. At the end of the growing season, purple-tinged seed pods open and the plant becomes a mass of silky down.

Distribution:

Fireweed is described as ranging from Siberia across northern North America. It is found in meadows and forests, on river bars, and in burned areas.

Medicinal uses:

According to Lantis, Alaskan Natives use fireweed tea for stomachaches. The tea from the leaves is stronger than chamomile tea and good for restlessness. (Hall)

Dried fireweed roots (gathered after the plant has dried) can be mixed with grease and spread on infected sores or bites.

Culinary uses:

Fireweed marrow, or pith, is eaten by Natives; the leaves are used for a tea called *kapor* in Russia. (Lewis) The root is eaten raw by Siberian Eskimos. Young tender greens are good in salads or cooked.

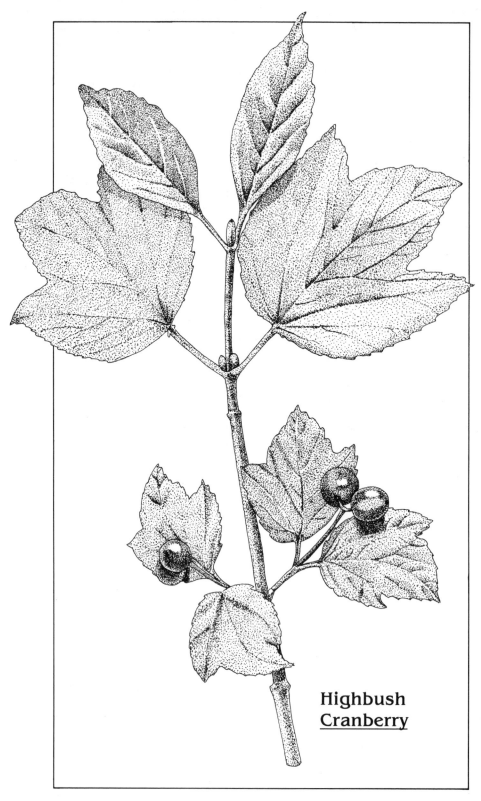

**Highbush
Cranberry**

34

Highbush Cranberry

Viburnum edule
(Caprifoliaceae)

Description:

The sour berries of this shrub hang on their stems, three to seven in a bunch, with all of the individual stems radiating from a single point. This characteristic distinguishes highbush cranberry from the poisonous baneberry, whose berries are attached to the upright main stem of the plant in an alternate manner.

Each berry of *Viburnum edule* contains one large flat stone; the interior flesh is red. The shrub grows up to 6 feet (2 meters) high with several to many stems. Its leaves are rounded, toothed, and shallowly three-lobed with a rounded base. The flower clusters are white cymes. Each flower has five stamens inserted on the corolla opposite the five lobes.

Distribution:

Highbush cranberry grows in woods and thickets all over Alaska except for the Aleutians, the Arctic and western coast.

Constituents:

In *The Merck Index,* dried bark of *V. opulus* (a relative of *V. edule)* is listed as containing viburnin, bitter resin, tannin, and sugar. It also has citric, malic, oxalic, and valeric acids. Nicholson et al. describes the antispasmodic effects of viopudial, a new non-alkaloidal material isolated from *V. opulus* bark.

Medicinal uses:

Its other common name of "cramp bark" and its listing in the therapeutic category "antispasmodic" in *The Merck Index* both indicate that highbush cranberry is a natural source of muscle relaxant. I have tried it as a treatment for menstrual cramps many times and it has always worked: use one cup of tea made from a handful of bark shavings.

A highbush cranberry leaf decoction has been used for sore throats. (Smith) The bark of the highbush cranberry is Tanaina medicine for stomach trouble; these natives boil and drink the tea. Upper Cook Inlet people say it is good for colds, sore throat, and laryngitis. (Kari)

In 1971, Nicholson, Darby and Jarboe reported the isolation of viopudial from *V. opulus*. Viopudial produced bradycardia, hypotension, and some decrease in myocardial contractility. Experimental evidence indicates viopudial's action is partly due to its effects on cholinesterase. Viopudial was relatively weak when compared to the known potent inhibitor physostigmine.

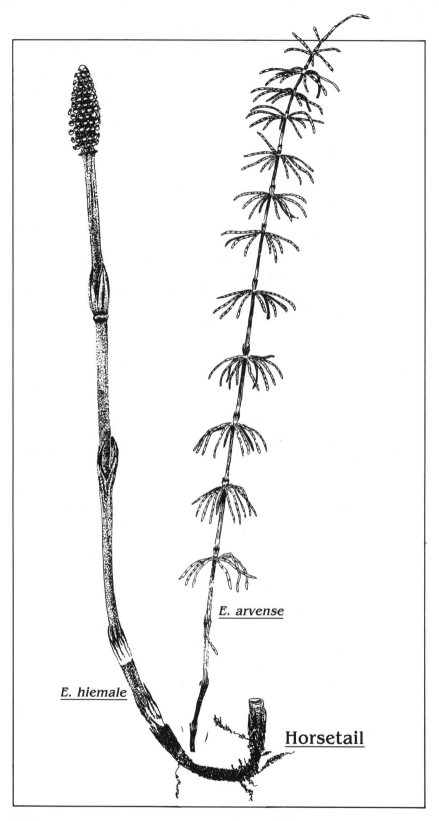

E. arvense

E. hiemale

Horsetail

Horsetail

Equisetum species
(Equisetaceae)

Description:

The round stems of horsetails are rough in texture, grasslike or rushlike, and grooved or furrowed. The ridges between the grooves are provided with almost-invisible spikes. The most obvious characteristic of the stems of horsetails, however, is the fact they are jointed (the plant is sometimes called joint-grass). At each joint the lower part wraps around the outside of the upper part.

Horsetails have no true flowers or seeds, but produce spores borne in fruiting bodies that somehat resemble the tips of asparagus stalks and soon wither away in some species. Some horsetails have no branches at all *(Equisetum hiemale)*; some species have whorls of needlelike, grooved leaves branching at the nodes *(E. arvense)*; and in one species *(E. sylvaticum)* the branches are again branched. Part of the "root" of the plant is swollen and round.

Distribution:

Horsetails are an extremely common ground cover in moist places.

Constituents:

These plants contain silica, aconitic acid, equisitine, starch, several fatty acids, and even some nicotine. (Spoerke) Thiaminase is an enzyme, toxic because it destroys thiamine. (Lewis)

Medicinal uses:

In the words of Simmonite-Culpeper, "The decoction, taken in wine, provokes urine, and helps the stone and strangury." Thus the horsetail has been used to promote the flow of urine if the flow is slow. In fact, *Equisetum* is one of the herbal diuretics. (Lewis) It has also been used for the treatment of diarrhea. (Simmonite-Culpeper and Spoerke)

In Washington state, Quileute swimmers rub themselves with *E. hiemale* to feel strong. The Cowlitz break up the stalks of the horsetail species, boil them, and wash hair infested with vermin in this water. The Quinault boil horsetails with willow leaves for a girl whose menstrual period is not regular, or use the juice for sore eyes. (Gunther) Horsetail has been used to cure internal wounds of the bowels, to treat obesity and dropsy, and to dissolve bladder stones. (Levy)

A Barefoot Doctor's Manual includes *E. hiemale* as one of the Chinese medicinal plants useful for conjunctivitis, inflammation of the lacrimal ducts, influenza and colds, dysentery, edema, hematuria, blood in stools, and menorrhagia.

Dr. John Christopher, a very popular herbalist whose herbal remedies are widely acclaimed, has decades of experience as a healer and teacher of herbalism. Some of his ideas are not accepted by the mainstream of modern medicine and science; his use of horsetail is a case in point. He uses it in his "bone, flesh, and cartilage" formula because of its silica, which, according to the theory of biological transmutation, is converted to calcium in the body and thus made available for use in growth, maintenance, and repair of skin and bones. This theory has a faintly alchemical tone to it, reminiscent of attempts to convert lead into gold. Of course, now it *is* actually possible to convert lead into gold, although the method is not economically feasible. So there may eventually be wide acceptance of the currently disreputable theory of biological transmutation, just as there have been "discoveries" by science of the validity of other practices from folklore, such as the use of quinine for malaria, cinchona for gout, rauwolfia for a tranquilizer, and penicillin in the form of bread mold for infections.

Warning:

While the external use of horsetail is probably safe, and it is used for healing and drying sores as a poultice, pack, or ointment, continued internal use is not a good idea. Although shoots are eaten when young, horsetail should not be consumed raw because of the toxic thiaminase (which is, however, destroyed by cooking). And persons with hypertensive disease and or other cardiovascular problems should not consider using it.

Iris

Iris setosa
(Iridaceae)

Description:

The iris, or blue flag, is a large, beautiful, lilylike violet-colored flower with markings of other colors. In all irises the mature flower consists of three large, drooping, petal-like sepals (the falls) outside of three erect, petaloid stigmas that arch gracefully over the stamen. The sword-shaped, erect leaves overlap. The stem is stout, straight, almost circular, and sometimes branching. The fruits — two rows of flat seeds — form in an oblong capsule, and the root is a cylindrical rhizome, often branched, and compressed toward the larger end where there are a cup-shaped scar and numerous rings of scars.

Distribution:

The distribution of *Iris setosa* is around the north Pacific. (Hulten) The genus *Iris* is a Northern Hemisphere group. (Gray)

Constituents:

Iris versicolor and *I. caroliniana,* two other North American species of iris, contain iridin, isophthalic acid, a camphoraceous substance, gum tannin,

Iris

39

sugar, and oil, according to *The Merck Index.* The rhizome of *I. versicolor* contains starch, gum tannin, volatile oil, acrid resinous matter, isophthalic acid, traces of salicylic acid, and possibly alkaloid. Its medicinal use is due to an oleoresin. (Grieve)

Medicinal uses:

Positive reports on the use of *Iris* include that of Smith, who describes the root infusion as a laxative. De Laguna says it is a magic plant to the Saint Elias people. In Angoon they make tea with iris roots; others use the whole plant. Anderson (cited in Oswalt) reports the seeds of iris are used as coffee (see Culinary uses, following).

Iris versicolor is described by Christopher as a cathartic, diuretic, stimulant, emetic, antisyphilitic, resolvent, sialogogue, anthelmintic, hepatic, and purgative. Iris is a powerful liver stimulant, said to be equal to mandrake root and less irritating. It clears the bile ducts of catarrhal obstructions, is beneficial to the secretive glands of the intestines, stimulates the flow of saliva, and is healing to the lymphatic system.

The medicinal action of *I. versicolor,* according to Grieve, is that iridin acts powerfully on the liver. Fresh iris is quite acrid, and if employed internally produces nausea, vomiting, purging and colicky pains. The dried root is less acrid and is employed as an emetic, diuretic and cathartic. The oleoresin in the root is purgative to the liver, and useful for bilious sickness, in small doses.

"It is chiefly used for its alternative properties, being a useful purgative in disorders of the liver and duodenum, and is an ingredient in many compounds for purifying the blood. It acts as a stimulant to the liver and intestinal glands and is used in constipation and biliousness, and is believed by some to be a hepatic stimulant second only to podophyllin, but if given in full doses it may occasion considerable nausea and severe prostration." (Grieve)

Culinary uses:

In spite of reports that iris root is poisonous (see Warning, following) I tried some tea from the seeds of the *Iris* in Fairbanks. The seeds are very easy to gather after the snow falls — they fall out of the capsule when you tip it. A handful was gathered in a minute or two.

A French chemist cited in Grieve, who discovered that the seeds produce a beverage similar to coffee and even superior to it in flavor, adds that the seeds must be free from the friable skin that envelops them and must be well-roasted before using. I did not agree. The unroasted seeds of *I. setosa* (he was using *I. psendacorus)* produced a delicious brew. Because the skin of the seed was not friable enough to be easily removed, I ground it up along with the whole seeds in a coffee grinder. (The seeds were too tough and resilient to be cracked in a mortar and pestle.)

Warning:

White reports that iris root is poisonous, and Lewis says that it is a poisonous glycoside. When ingested it may cause dermatitis due to allergenic reactions. (Moulton, 1979)

Juniper

Juniper
Juniperus communis
(Cupressaceae)

Description:
A member of the Cypress family, juniper in Alaska grows as a shrub with prostrate form and extremely sharp-pointed needlelike leaves. (There is a rare species in southeastern Alaska, *Juniperus horizontalis* or creeping juniper.) The fruits of *J. communis* are berries, green the first year ripening to black in the second or third year.

41

Distribution:

This species is the most widely distributed conifer in the world and the most widespread tree species in the North Temperate zone.

Constituents:

Volatile oil is found in highest concentration in full-grown but unripe berries. You should obtain 0.5 ml. of juniper oil from each 100 grams of berries. The riper the berries, the more of the volatile oil has been converted to resin. (Spoerke) The volatile oil has been called oil of sabinal. It is obtained by steam distillation and contains 50% alcohols, primarily 1-terpinen-4-ol, alphapinene, camphene, and cadinene. (Spoerke) The fruit also contains resin, sugar, tannin, and a flavone of glycoside. *The Merck Index* cites juniperin as a constituent.

Medicinal uses:

Usually the berries are used, chewed raw or steeped into tea. The major use of juniper by herbalists has been for its stimulant effect on the genito-urinary tract. (Juniper oil is one of the ingredients in many patent remedies.) Besides its diuretic effect, juniper is used for flatulance, to increase heat, to stimulate the stomach, to strengthen the brain, for rheumatic pains, and for problems in the back and chest.

In Alaska, the Tanainas like to burn the needles on top of a hot wood stove as an incense. They say smelling them is good medicine for a cold. I have tried this many times, as I find the odor of smoking juniper needles delightful.

Kari also reports that Tanainas boil branches of juniper in water and drink the tea for colds, sore throat, and tuberculosis. They say it helps a person who has a hard time urinating.

In the Interior of Alaska, traditional medical practice included two uses of juniper for colds: boil berries with leaves and drink ½ cup; or boil squashed berries, strain, and drink 1 cup. (Hall)

The leaves were used by the Hopi as a decoction for a laxative or as a drink for women who wanted daughters. Juniper is chewed or given as a tea to women after childbirth, and for twenty days afterward the mother's food is prepared with the addition of some juniper leaves. The mother's body is washed in juniper tea and the newborn baby is rubbed with ashes from burned juniper. When men return from burying a corpse they wash themselves outside the house in water containing a branch of juniper. (Leek)

Goodman and Gilman's *The Pharmacologic Basis of Therapeutics* (p. 970) mentions a *Juniper tar,* U.S.P., which is used as a 1 to 5% ointment, 4% shampoo, or 35% bath emulsion. "The medicinal tars are sometimes considered to be antiseptic because of various phenolic components. However, whatever efficacy they have in their uses mainly results from a mild irritant effect. The tars are used in the treatment of diseases of the skin such as psoriasis and eczema-dermatitis."

In herbology of cold remedies many gymnosperms have been used as teas or inhalants to relieve cold symptoms. "The Dakotas, Omahas, Poncas, and Pawnees burned the twigs of *Juniperus virginians* and inhaled the smoke for colds and also used them in vapor baths; a decoction of the boiled cones and leaves was used by the Plains tribes for treating coughs." (Lewis and E. Lewis)

Labrador Tea

Labrador Tea

Ledum palustre

(Ericaceae)

Description:

The narrow leaves of Labrador tea are evergreen. Shiny dark green and leathery above, slightly woolly below, they have smooth margins that curl under. The leaf undersides are white in young leaves, cinnamon-brown in older ones. The thin stems of this shrub are hairy when young. The white flowers form umbel-like clusters. A sweet, spicy aroma is noticeable when walking through these knee-high bushes.

Dwarf Labrador tea is very similar in all respects to the common form except its leaves and overall size are smaller.

43

Distribution:

This species is holarctic — distributed all around the northern regions of the planet. It is one of the most common shrubs.

Constituents:

The tannin is called leditannic acid, and there are gallic acid (a bitter substance), wax, resin, salts, and ascorbic acid. Be warned that Labrador tea contains ledol, a poisonous substance causing cramps and paralysis. Grieve says the plant also contains a stearopten, valeric and volatile acids, ericolin, and ericinol.

Medicinal uses:

Labrador tea is common, widespread, and always available in northern climates where non-evergreen leaves are obtainable only during a short growing season. The plant also has a pleasant aromatic scent, lending a spicy fragrance to a tea. For these reasons it is perhaps no wonder this plant is mentioned in such a large number of ethnobotanical reports and herbal compendia. A small amount added to black tea does add a spicy aroma. It is used in this way on Nelson Island (Ager and Ager) or mixed with willow leaf tea. (Lantis) Several authors (Tobe, de Laguna, Grieve) indicate that Labrador tea is good for colds. The leaf tea is cathartic if it is strong enough to be orange-colored. The ascorbic acid content is second only to rosehips. (Lantis)

The leaves and occasionally the twigs and flowers are used to make tea. The method of gathering and preparing the leaves varies greatly. I usually use them fresh and prefer the older leaves, but some herbalists prefer the young leaves and dry them. Be sure to dry them slowly and carefully so they do not turn black.

Some Native peoples in western Canada steam the leaves until they turn dark brown and place the rhizomes of the licorice fern *(Polypodium glycyrrhiza)* in with them for flavor. (Turner & Szczawinski) Hall says to drink it for rheumatism.

Other uses:

There are ceremonial uses for Labrador tea; one is to turn a stalk and throw it out the door if a child is ill or if you want to get rid of ghosts. (Oswalt)

Bees are much attracted to the flowers, but animals do not browse much on the plants, which are slightly poisonous. Leaves strewn among clothes impart a fragrance pleasant to most people.

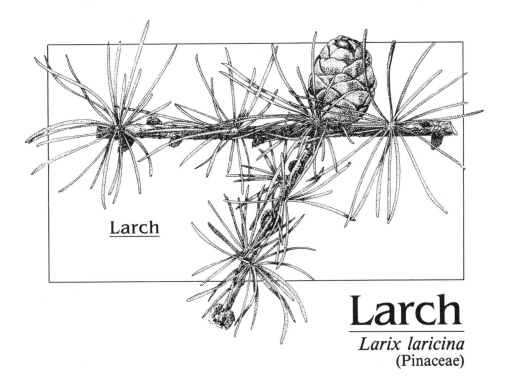

Larch

Larch
Larix laricina
(Pinaceae)

Description:

A small tree, larch grows to about 33 feet (10 meters) tall, with brownish bark, horizontal branches, and a thin crown. The leaves, deciduous needles, form fascicles of ten to twenty on short spur branches. The branches are long, very narrow, slender, and flexible, three-angled and blue-green. Before falling in early autumn they turn yellow. The twigs are long, stout, dull tan, and hairless, with many short stout spur twigs. Winter buds are small and round, about ½ inch (2 mm.) long and covered by many short, pointed, overlapping scales. Rounded, upright female cones measure ⅖ to ¾ inch (1 to 1.5 cm.) long. Some of the tree's common names are Alaska larch, eastern larch, hackamatack, and tamarack.

Distribution:

The larch is found across northern North America.

Medicinal uses:

Larch bark tea is laxative, tonic, diuretic, alterative. It is useful in obstructions of the liver, rheumatism, jaundice, and some cutaneous diseases. (Grieve) She also reports the use of the leaves as a decoction for piles, haemoptysis, menorrhagia, diarrhea, and dysentery.

The inner bark, made into tea, is good for bleeding of any kind, hemorrhoids, excessive menstruation, and as a tonic to the liver and spleen. (Kloss) Tobe says to mix with spearmint, juniper, and horseradish in a bark decoction for liver and skin disorders, rheumatism, as a laxative, and externally for piles, itching, menorrhagia and dysmenorrhea.

Culinary use:

In the spring the buds can be eaten raw or cooked. They are quite sweet.

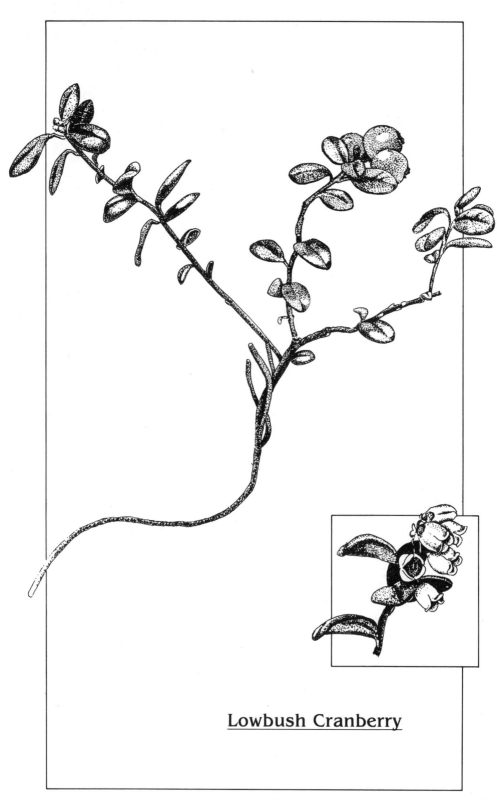

Lowbush Cranberry

Lowbush Cranberry

Vaccinium vitis-idaea
(Ericaceae)

Description:

This evergreen subshrub is creeping and mat-forming. Its leaves are ovals ⅜ to ¾ inch (10 to 20 mm.) wide, shiny above, light green beneath, and spotted with short, stiff brown hairs. (Brown spots distinguish this plant from uva-ursi.) Leaf edges roll under slightly. One to several flowers nod on short stalks, less than ⅛ inch (1 to 2 mm.) long, at the ends of the twigs. The corolla is pink and bell-shaped. The fruit, a bright red, very abundant berry, has white interior flesh. Berries are best if picked after the first frost or during the winter and spring.

Distribution:

Lowbush cranberry, also called lingonberry and mountain cranberry, is found throughout Alaska and the rest of North America, northern Europe, and Asia.

Constituents:

According to *The Merck Index, Vaccinium vitis-idaea* contains vaccinin.

Medicinal uses:

Chew lowbush cranberries for a sore throat. For headaches, swelling, and sore throats, including tonsil troubles, just heat the berries, wrap them in a cloth, and put them as a hot pack on the sick place. (Kari) Hall says to munch on berries to relieve an upset stomach and, for measles, to boil the cranberries, rub on the measles rash, and cover.

Plantain

Plantain

Plantago major
(Plantaginaceae)

Description:

At the base of each plantain is a large rosette of strongly ribbed leaves; these leaves are broad, ovate, blunt, and abruptly contracted at the base with winged petioles.

Beginning in the second year, each plant produces one to several cylindrical spikes, ½ to 1 foot (15 to 30 cm.) tall. These inflorescenses, leafless along the stem, bear many small, inconspicuous greenish-white flowers. Each flower has four overlapping, persistent sepals with dry membranous margins, a flat and circular corolla (that soon withers) over the capsule, four stamens, and a compound pistil with a long threadlike style. The stigma matures well before the stamens shed pollen, thus ensuring cross-pollination.

Plantain fruits, small ovoid capsules, contain eight to eighteen angular seeds with a netted surface pattern.

Distribution:

Plantain is found worldwide, on roadsides and in waste places.

Constituents:

Plantago contains potash salts.

Medicinal uses:

Plantain leaves, roots, flower spikes and seeds are all used medicinally.

Both the leaves and the roots manifest moderately diffuse and stimulating alterative effects within the circulatory system. These beneficially influence the glandular system, with marked healing to affected lymph and epidermal areas in scrofulous and skin diseases.

The fresh leaves, pounded into a paste, are used to check bleeding. The tea may be applied to skin irritations and diseases. For hemorrhoids, it has been injected or applied externally with a piece of gauze. Mashed green plantain leaves are applied as a poultice for bites, boils, carbuncles, and tumors.

Plantain is an effective treatment for poisonous bites and stings, since the poison of fresh stings is extracted rapidly. It is very useful for easing pain and healing problems in the lower intestinal tract. (Christopher)

As an astringent, plantain contracts tissues and has been used for excessive flow in menstruation. It is used as a douche for leucorrhea and syphilis, and as a tea for diarrhea, kidney problems, and bladder trouble. It is mixed with sourdock as a wash for itching skin, ringworm, or running sores. *A Barefoot Doctor's Manual* lists it as diuretic and cooling.

Lucas tells us the Romans called this plant waybread. Dioscorides recommended it for leg ulcers; Pliny says a book was written on it. Saxons and Iroquois both bound crushed leaves around the head for headaches.

Culinary uses:

The very young leaves are packed with vitamins A and C and can be eaten as a salad green or steamed as a cooked vegetable.

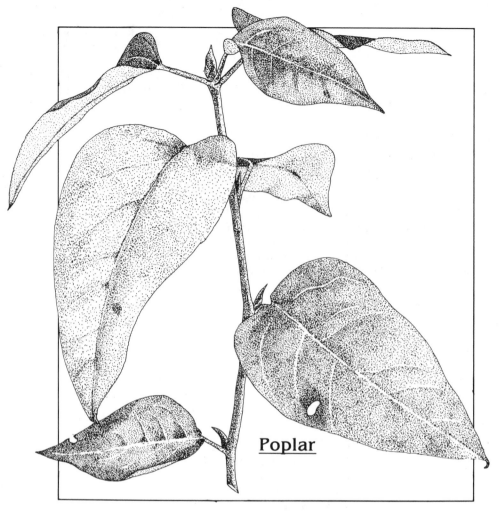

Poplar

Poplar or Cottonwood

Populus balsamifera
(Salicaceae)

Description:

A medium-sized tree, usually 30 to 50 feet (9 to 15 meters) tall, poplar has a straight trunk with bark that acquires deep diamond-pattern ridges in older trees. The winter buds are the largest, most aromatic, and most sticky-resinous of the deciduous trees. These buds are about 1 inch (2.5 cm.) long, pointed, and covered with shiny brown scales. They release a pungent balsam odor. The twigs are reddish brown and the leaves, with slender petioles, have sharply pointed blades, rounded bases, and many small, rounded teeth along the margins. The flowers form long, hanging catkins.

Distribution:

Poplar, which can hybridize with quaking aspen, is distributed throughout Alaska (except the coast) and across North America. It is especially common on well-drained gravel along rivers and occurs farther north than any other tree species.

Constituents:

The buds of *Populus candicans,* a relative, contain a balsamlike resin, a yellow volatile oil (primarily humulene), gallic acid, malic acid, mannite, chrysin, tectochrysin, a fixed oil, and two glycosides, salicin and populin. (Populin is salicin benzoate.) (Spoerke)

The antibiotic trichocarpin, active against fungi, comes from the bark of *P. candicans;* bisabolol, active against tuberculosis bacilli, comes from young shoots of another relative, *P. tacamahaca.* (Lewis and Elvin-Lewis)

Medicinal Uses:

The salicin in poplar buds has some action against fever and pain; the resin yields a terpene to which many therapeutic effects are attributed. (Spoerke)

Poplar buds, used externally, are mild as a counterirritant or expectorant. Cottonwood bud tea is used for colds by Interior Alaskans. (Hall) Salve from the winter buds of *P. balsamifera* is simple to make. (See balm of Gilead procedure, following.) Lime villagers use this as medicine for a sore, rash, or frostbite. (Kari) The Menominees and Pillager Ojibwas rubbed the ointment on the nostrils or put it up the nose, allowing the balsamic vapors to course through the respiratory passages to relieve congestion from colds, catarrh, and bronchitis. They boiled the buds in mutton or bear tallow. (Lewis and Elvin-Lewis) I have used balm of Gilead ointment in this way myself.

The term "balm of Gilead" is applied to an ointment made from the buds of the poplar tree as well as from a totally different tree from Arabia. Collecting the buds is easiest at temperatures just below freezing, when they are frozen enough that they are not sticky and do not adhere to the fingers, but not so cold that hands get chilled. It is the terminal bud that I gather.

The procedure for making balm of Gilead is simply to cover the winter buds with oil, heat them, then strain and use the liquid. Spring buds are all right to use until they have begun to open, when the amount of aromatic oil-soluble resin is relatively less compared to the amount of water and water-soluble materials.

The kind of oil to use is a matter of personal preference. One book recommends wolverine fat, but usually I use a fine vegetable oil and then add some beeswax to the point where it does not run, but has the thickness of a lotion. Vaseline or cocoa butter work very well also, as would lard, lanolin, or solid shortenings. The heat should be low enough to warm but not boil the oil, or the buds might burn. To maximize the extraction, the amount of time to heat can be extended for days if the pot is on the back of the wood stove, but twenty minutes would be a reasonable time on a low fire.

Christopher lists the following herbs to add to the balm to increase its efficiency: anise root or sweet cicely, chickweed, coltsfoot, horehound, hyssop, licorice, lobelia or red sage.

Poplar bark is cathartic, tonic, stimulant, diuretic, anti-scorbutic, stomachic, resolvent, discutient, alternative, expectorant. The buds are stimulant, tonic, diuretic, expectorant, nephritic, demulcent, cathartic, peristaltic, and nutritive. (Christopher)

Raspberry

Raspberry

Rubus idaeus
(Rosaceae)

Description:

The fruit of the raspberry bush is so well-known it needs no description. The genus *Rubus* includes many other species having the characteristic raspberry or blackberry-type fruit, with many tiny druplets clustered together to make a compound fleshy berry. These species all bear edible berries and most can be used for high-quality leaf tea, although the red raspberry is the one most widely used for this purpose. The leaves of *Rubus idaeus* are compound with three to five leaflets, the flowers are white and clustered, and the stem is armed with small thorns.

52

Distribution:

The wild raspberry, from which the cultivated variety has been derived, occurs naturally all across northern North America and Eurasia except for the high Arctic and tundra regions. This shrub tends to be a successional species; thus it is especially common during the first ten years after an area has been disturbed. Berms and roadsides are likely to be good places to harvest raspberries.

Constituents:

The berry offers vitamins A, B1, B2, calcium, phosphorus, and iron, in addition to a fruit sugar, a fragrant volatile oil, citric and malic acids, pectin, coloring matter, and water. The leaf is high in citrate of iron and some tannins.

Medicinal uses:

Dried raspberry leaves make a safe, pleasant-tasting tea that has been highly regarded by both European and Indian peoples for stomach complaints and as an effective remedy against diarrhea and dysentery, presumably as a result of its astringent action.

A sure remedy for vomiting and/or diarrhea is to let a strong solution of tea cool to lukewarm and sip slowly.

The tea is recommended as a general tonic for pregnant women. I have known several who drank it freely during their pregnancy and labor and while nursing, under the impression that it served to prevent or alleviate the nausea of morning sickness, promote the general well-being and health of the reproductive system, and prevent premature labor pains. It has been shown in animals to relax the smooth muscles of the uterus and intestine (Spoerke), so it might not be advisable to drink the tea during labor. In the post-delivery period, the tea is said to relieve after-pains. (Christopher)

Externally the tea was used as a wash for sore mouths, wounds or ulcers, and as a gargle for sore throats, especially for young children.

A related plant, R. spectabilis, the salmonberry, has bark and leaves containing an astringent quality. They are pounded or chewed and used as a poultice on burns or wounds. (Gunther)

Culinary use:

The delicious flavor of a fresh red raspberry is a rare treat, more difficult to buy in some parts of the country than a Rolls-Royce.

Warning:

Although there is no evidence of toxic effects from drinking raspberry tea, the leaves do contain tannins, so continued use might introduce the possibility of bowel or kidney irritation. (Spoerke) Also, the leaves must be completely dried before using — partially wilted leaves can be harmful. (Turner & Szczawinski) Old dead leaves or twigs are safe so long as they are dry.

My method of drying the leaves is to cut whole stalks and hang them to dry. The brittle leaves can then be removed easily by breaking and crumbling. (The under-dried green leaves are not easily removed by pulling off the stalk.)

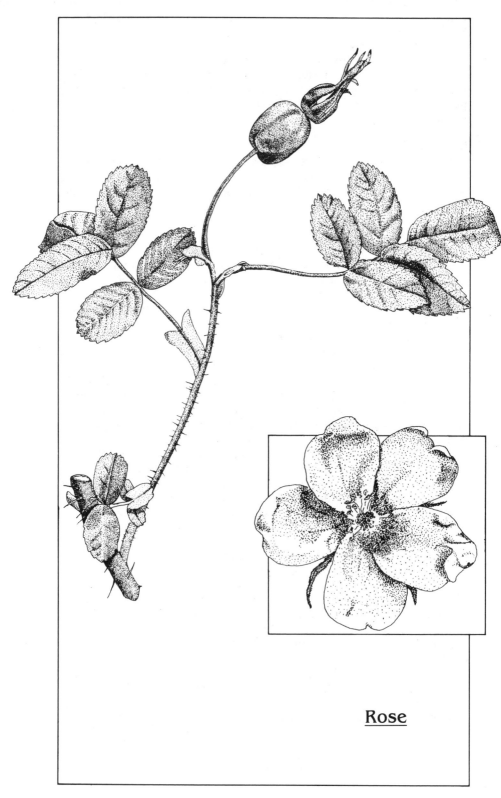

<u>Rose</u>

54

Rose

Rosa acicularis
(Rosaceae)

Description:

A small shrub with bristly and prickly stems and branches, this rose has three to seven opaque, odd-pinnate, serrate leaflets, hairy on the undersides, with puberulent and often glandular rachis. The solitary flowers, about 1½ to 2½ inches (4 to 6 cm.) in diameter, display five rose-colored petals. The sepals are erect in fruit, glandular on the back. The subglobose hips contract to the neck below the sepals.

Distribution:

Roses are found in woods, heaths, tundra bogs, and thickets. In Denali National Park the plant occurs to about 1,100 meters. It is described in Siberia.

Rosa acicularis is the most widely distributed of the three species of roses in Alaska. The other two similar species are *R. woodsii,* which escaped from cultivation, and *R. nutkana,* which occurs in the southeastern and southern coastal regions and hybridizes with *R. acicularis* where their ranges overlap.

Medicinal uses:

Rosa acicularis is an excellent source of vitamin C. Its hips and leaves are antiscorbutic; the bark is emetic.

For sore eyes, wash with juice made by soaking the flowers in hot water.

Stems and branches are used for colds, fever, stomach trouble, weak blood, and menstrual pains. Burn the thorns off, then break up the twigs and boil them. For a drink to cause vomiting, soak the bark in hot water until the solution is very strong. (Kari)

The Skagit of Washington state make a sore throat medicine by boiling the roots of *R. nutkana* with sugar. The Cowlitz use rose leaf tea for bathing a baby. (Gunther)

Culinary use:

The flowers are good to eat, but the white pip at the base should be cut out.

**Shepherd's
<u>Purse</u>**

Shepherd's Purse

Capsella bursa-pastoris
(Cruciferae)

Description:
The typical mustard family fruit is, in the case of shepherd's purse, a distinctive heart-shaped capsule, each one on its own stem at right angles to the main stalk. This seed pod resembles the common leather purse of pastoral peoples, thus the name "bursa-pastoris," which means shepherd's purse. The small white flowers bloom among the ripening seed pods on an erect stalk bearing a few arrow-shaped leaves. Most of the deeply dissected and narrow leaves grow in a basal rosette. The stem leaves are clasping, i.e., their bases wrap around the stem.

Distribution:
This plant is worldwide, a cosmopolitan weed — one of the most common and persistent garden weeds. I have never seen it in wilderness areas of undisturbed vegetation.

Constituents:
Shepherd's purse contains vitamin K and a tannate, an alkaloid, bursine, and bursinic acid. There is a volatile oil, similar to oil of mustard, and 6% of a soft resin. (Grieve)

Medicinal uses:
The very widespread use of shepherd's purse against bleeding of all kinds may be associated with its high vitamin K content. It has been employed internally as a tea and externally as a poultice on wounds. The herb has also been shown to have anti-ulcer and anti-inflammatory actions. Other traditional functions are against fever and diarrhea.

To corroborate the traditional medical uses of shepherd's purse, Kuroda et.al. have investigated the pharmacological actions of the alcohol extract of the herb. It was found to decrease blood pressure and increase peripheral blood flow. The extract contracted smooth muscles such as the small intestine, tracheal muscle, aorta, and uterus. It prolonged the sleeping time induced by sodium hexobarbitone and prevented the induction of hepatomas. It showed low toxicity in mice. The growth of Ehrlich tumor was clearly suppressed by injections of the extract.

Culinary use:
The greens are quite palatable as a potherb.

57

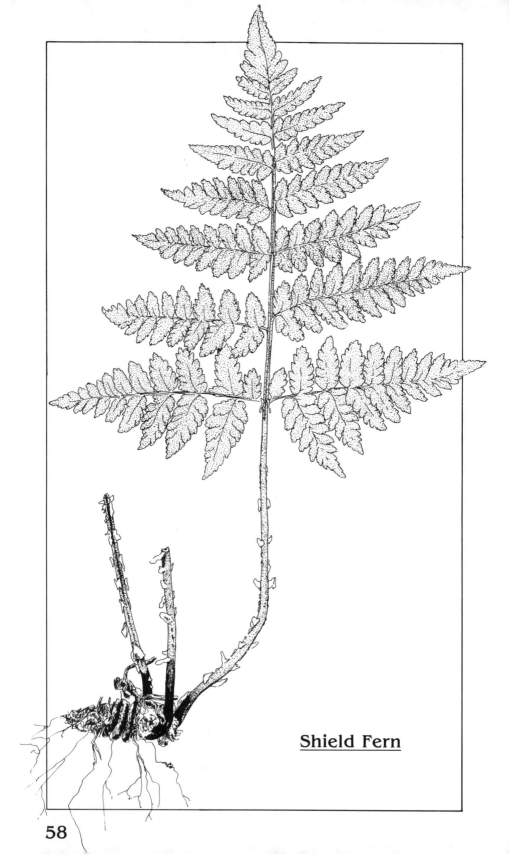

Shield Fern

58

Shield Fern

Dryopteris dilatata
(Aspidiaceae)

Description:

Ferns have no true stems. Their leaves are fronds, 2 to 4 feet (.7 to 1.5 meters) long, wide and spreading. Pinnae are oblong and rounded, with their edges slightly notched and their surfaces somewhat furrowed. The sori, or clusters of spore cases, appear on the upper half of the frond at the back of the pinnules, in round masses toward the base of the segments. The fruits are spores, dustlike and almost invisible. The rhizome is horizontal and creeping. The crown forms a brown, tangled mass.

Distribution:

The shield fern grows in most wooded, moist areas throughout Alaska. *Dryopteris fragrans* in the Interior is smaller.

Medicinal uses:

The therapeutic action of the rhizome is anthelmintic, astringent, tonic, and vulnerary. Lantis calls it a comforting tea for the gut, used by Alaskan Natives. Kari reports that the Lime villagers say to boil it hard, then wash your eyes with the cooled tea or drink it for kidney trouble and breathing problems such as asthma.

Culinary use:

Curled fiddleheads (new fronds) are edible.

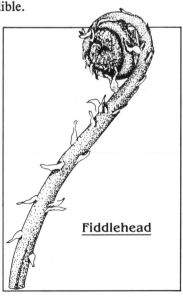

Fiddlehead

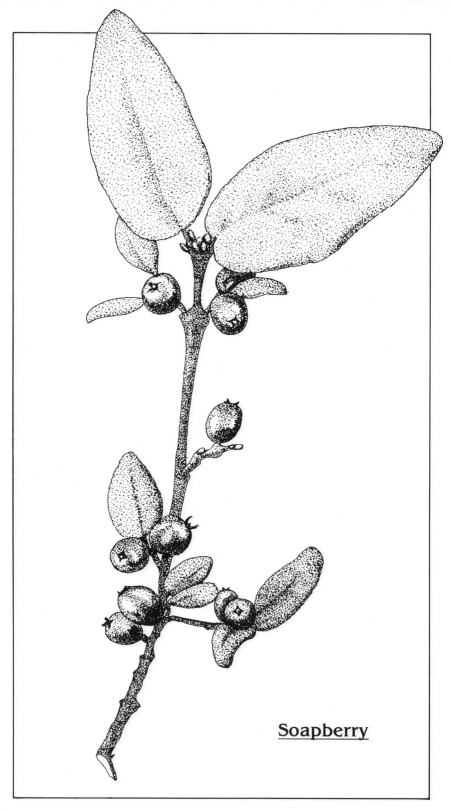

Soapberry

Soapberry
Shepherdia canadensis
(Elaeagnaceae)

Description:

This deciduous shrub reaches 3 to 6½ feet (90 cm. to 2 meters) tall. The twigs are gray and scaly with paired branches. Reddish-brown scales cover young twigs and buds.

Soapberry leaves are opposite, with short scaly petioles less than ⅛ inch (3 mm.) long and without stipules. The ovate blades, ½ to 2 inches (1.2 to 5 cm.) long, ¼ to 1 inch (0.6 to 2.5 cm.) wide, are rounded or blunt on both ends, and not toothed on the edges. The upper sides of the leaves appear green and slightly hairy with scattered star-shaped hairs, while underneath they are densely covered with reddish-brown scales and silvery, star-shaped hairs.

The yellowish or brownish small flowers measure about ¼ inch (5 mm.) wide. These flowers, male and female on different plants (dioecious), bloom in short lateral spikes in the spring before the leaves form. The male flowers have a calyx of four spreading, scaly lobes and eight stamens. The fruits, elliptical, red or yellowish, and about ¼ inch (6 mm.) long, look nearly transparent. They are fleshy and edible, but almost tasteless and bitter.

Distribution:

Soapberry is described as occurring throughout Alaska in dry well-drained soil, especially near lakes and rivers.

Medicinal uses:

Kari says to make a tea with the stem and drink it for tuberculosis, or wash cuts and swellings with the tea.

Culinary uses:

Gathered in quantities, the fruits were eaten by Indians. Fruits pressed into cakes were smoked — the first sweet taste is replaced by a bitter taste (saponin), like quinine. Also, the fruits were mixed with sugar and water and beaten into an edible foam or froth that was used like whipped cream on desserts.

Sourdock

Sourdock

Rumex arcticus
(Polygonaceae)

Description:

The members of the genus *Rumex,* a widespread group with thirteen species represented in Alaska, are the docks and some of the sorrels. *Rumex arcticus* closely resembles the widely known *R. crispus,* called curled dock or yellow dock. *Rumex arcticus,* a perennial about 3 feet (1 meter) tall, has coarse-ribbed stems growing from a fleshy taproot. Its leaves are sometimes red-tinged, quite large (especially near the base of the stem), and long and narrow. A sheath wraps around the stem at the base of each leaf. The small flowers form panicles or clusters at the end of the stalk. The fruits are characteristic, three-angled like the flowers, with winglike outgrowths formed from the three inner sepals.

Distribution:

Rumex crispus, called curled dock due to the wavy margins of the leaves, is a native of Europe but has naturalized in most parts of the world, including interior and southcentral Alaska. *Rumex arcticus* is common in wet places throughout Alaska, adjacent Canada, and across Siberia.

Constituents:

The tannin from *R. hymenosepalus* root yields leucodelphinidin and leucopelargonidan, which have been employed as cancer chemotherapeutic drugs. (Lewis)

Medicinal uses:

Known as medicinal since ancient times, the curled dock or sourdock root has been used as a laxative, astringent, tonic, blood purifier, and alterative. European herbalists used sorrel or dock as a cooling agent for fevers, a bitter tonic appetizer, and as a "cordial to the heart."

The Upper Cook Inlet and Nondalton peoples used a tea made from sourdock root as a medicine for stomach and bladder trouble, for hangover, as a laxative, and, in a very strong decoction, as an emetic.

Roots were cooked, mashed with oil and put on cuts by the people of Mount Saint Elias. (de Laguna) And the identical use of dock is described by Euell Gibbons, whose directions for dock salve follow:

Boil fresh dock root in vinegar just to cover until the root is soft and most of the vinegar has boiled away. Cool until it can be handled, then put the root through a sieve or colander to remove fibers. Combine 1 part dock root to 2 parts petroleum jelly and work in a bit of dry sulfur. This salve has a reputation for curing an itch on man, and mange, saddle sores, and other external conditions on animals.

For relief from pain and itching, a home remedy is to apply the juice of *Rumex* species growing conveniently close to the stinging nettles *(Urtica* species) that caused the rash.

Internal use of the root of curled dock *(R. crispus)* is recommended by Christopher as a blood purifier in cases of poisoning with arsenic or heavy metals. He claims that iron compounds in the root promote excretion by chelation.

Other use:

Rumex hymenosepalus contains as much as 35% tannic acid. Its root has been used as a commercial tanning agent. (Szczawinski and Turner)

Warning:

The leaves of dock and sorrel (both *Rumex* and *Oxalis* species) and rhubarb contain soluble oxalates. Oxalic acid is the only organic acid of plants toxic to animals under natural conditions. (Smith) The soluble oxalates are sodium and potassium; calcium oxalate is insoluble. A small to moderate amount of the oxalates is not harmful, and the hazard can be somewhat mitigated by adding a food rich in calcium to a meal in which oxalate-rich greens are consumed. Thus the popularity of cream sauce with these greens makes nutritional sense.

There is also the unlikely hazard of allergic reactions and the chance that long-term consumption of large amounts of the tannin-rich roots might lead to irritation of the bowels or kidneys.

Sphagnum

Sphagnum

Sphagnum species
(Sphagnopsida)

Description:

The genus *Sphagnum* is in a class by itself. Also called peat moss or bog moss, sphagnum is usually wet and often grows in large hummocks. Individual plants measure just ¼ inch (7 mm.) long. The stem of each plant is not forked. The branches at the tip of the stem crowd into a ball called the capitulum, so that when you look down on the surface of the hummock each individual plant appears as a circle. Down within the hummock, the side branches of the individual plants are arranged in bunches called fascicles. The leaves on the branches form spirals and press close to the branch. Stem leaves differ in size and shape. Colors vary from red, green, and yellow to brown.

Distribution:

Sphagnum follows the sedges in bog succession and creates an acid environment in peat bogs, where it exerts an important effect on drainage, erosion and permafrost.

Medicinal uses:

This soft, acidic plant absorbs moisture better than a sponge. It is bactericidal, resistant to decay in herbaria, and safe to use against skin. Sphagnum was used extensively for surgical dressings during the first World War, and by mothers in Lapland and North America for infants' cradle linings, diapers, and toilet paper. (Crum, Oswalt) I have used it for my own infant and agree that it seems to prevent diaper rash.

Red sphagnum is good for a dressing on a bad cut or for sore eyes. (Lantis) It can also be used for menstrual pads. (Kari) Sphagnums help to combat infection; sometimes they are boiled or warmed on hot rocks first, then applied to an infected place and covered with a bandage. If you have to put a splint on a broken bone, sphagnum is the best moss to use for padding.

Other uses:

Sphagnum is often used for chinking log cabin walls. Indians of Guatemala believe the Christ Child was bedded in sphagnum.

Spruce

66

Spruce

Picea species
(Pinaceae)

Description:

Spruce trees are evergreens with needles borne singly. These square or more or less flattened needles have sharp tips. *Picea glauca* (white spruce) trees do not have hairs on their twigs, and the cones are longer and more oval in outline than black spruce cones. *Picea mariana* (black spruce) cones are roundish ovals; young twigs have fine hairs between the needles. *Picea sitchensis* (Sitka spruce), the largest tree of Alaska, is the state tree.

Distribution:

In the interior forests, white spruce is the most common tree. It occurs from near sea level to treeline at about 1,000 to 3,500 feet (305 to 607 meters). It grows best in well-drained soil on south-facing gentle slopes, and in sandy soil along the edges of lakes and rivers. (L. Viereck)

Black spruce is characteristic of north-facing slopes and lowlands underlaid by permafrost.

Sitka spruce is coastal in distribution.

Medicinal uses:

The pitch or gum from the spruce tree has been used for medicinal plasters. Pitch from both the European Norway spruce and the North American spruce has been used as medication for cuts and scratches. (Lust, Carroll) Spruce pitch's healing properties may be due to the fact that it keeps the wound clean, preventing infection during the natural healing process. The pitch should be clear and soft and it should be covered with a bandage. (Hall)

Hall lists some additional uses for spruce pitch: smear pitch on canvas, melt it in an oven, and use it to cover a sore back; put pitch on a large piece of cloth, put snow in the cloth, and wrap the cloth around the head to relieve a headache; mix pitch with just enough grease to spread it on infected sores or cuts; boil pitch and drink as much as you can for urinary problems. For blood poisoning, spread pitch on a cloth and wrap the infection site and red streak up as far as it goes; rub pitch on warts. If the pitch is a sticky white gum, boil it 5 to 10 minutes and drink for chest and head colds.

Uses for spruce needles from Hall are: boil the needles for one hour, strain, and wash with the solution to clear up hives or rashes; boil needles 5 to 10 minutes and take 2 to 3 teaspoons two to three times a day for a cold; drink 1 cup of spruce needle solution a day to purify the blood; boil needles constantly to rid the house of infections; boil needles and drink the solution, or dilute needle solution with water and sit in it, for urinary problems.

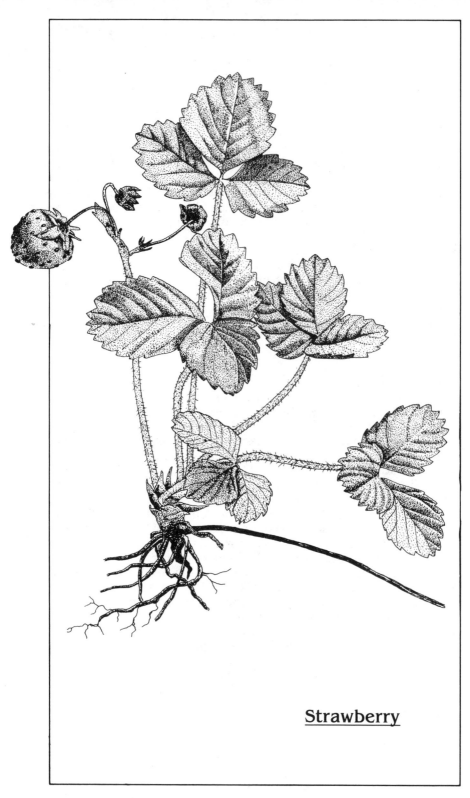

Strawberry

Strawberry

Fragaria virginiana and *F. chiloensis*
(Rosaceae)

Description:

The berries of the strawberry plant are so well-known they need no description. The leaves divide into three sharply toothed leaflets, with the terminal tooth smaller than the adjacent ones. The white flowers have five petals; the root is a thick rhizome. Runners or stolens are usually present.

Distribution:

Fragaria virginiana is coastal and *F. chiloensis* is restricted to the Interior. Both species hybridize with domestic varieties of *F. vesca*.

Constituents:

The leaves are known to contain vitamin C, catechins, and leucoanthocyanin.

Medicinal uses:

The ascorbic acid (vitamin C) in strawberry leaves and berries is both astringent and antiscorbutic. Strawberries' reputation for tightening loose teeth is probably due to their ascorbic acid — loosening of the teeth is one of the symptoms of vitamin C deficiency and scurvy. The alleged power of strawberries to dissolve tooth plaque is questionable, however, because any acid strong enough to dissolve the calcareous deposits on teeth is also strong enough to damage the tooth itself. (Nancy Georgell, personal communication) But the nutritional effect of a large amount of fruit in the diet may indeed be able to delay the formation of plaque.

According to Spoerke, catechins (found in strawberry leaves) are protein precipitants and astringents. The d-catechin is thought to inhibit production of histamine. It has little therapeutic action on its own, but seems to potentiate antihistamine drugs if used with them.

Strawberry roots are diuretic. (Densmore)

Strawberry leaves and roots, boiled in wine and water, are described as a remedy for diarrhea by Simmonite-Culpeper of medieval England. In their words, "The same if drank, stays the bloody flux and womens' courses, and helps the swelling of the spleen." The same plant used for the same purpose appears in the pharmacopeia of the Indians of western Washington (Gunther, Lewis); one wonders whether this is a case of independent discovery of a remedy or whether there was cultural transmission one way or the other. Also, the Chippewas used strawberry roots for a tea to give to children with stomach upsets. (Densmore)

Sweet Gale

Sweet Gale

Myrica gale
(Myricaceae)

Description:

This shrub grows up to 3 feet (1 meter) tall, with odorous resin-dots on its stem. Branches are reddish brown. The leaves are deciduous, oblanceolate, ⅗ to 2 inches (1 to 6 cm.) long, somewhat serrate toward the apex, and grayish green. The flowers, inconspicuous stiff spikes, develop before the leaves.

Distribution:

Sweet gale grows in bogs, swamps, shallow water, and along streams. It occurs south of the 67th parallel in North America and in Eurasia.

Constituents:

Von Schantz and Kapetanidis published an analysis of the composition of the essential oil of leaves from a population of *Myrica gale* from Finland. This oil is composed of about 130 constituents, of which 49 are present in a concentration greater than 0.1%.

Medicinal uses:

Sweet gale is used as a wash for boils and pimples and as a steam bath switch. (Kari)

Myrica cerifera is the bayberry of the east coast of the United States. Its root bark is astringent, tonic, alterative, cholagogue, diuretic, and aromatic. Bayberry is a powerful stimulant, cleansing and restoring the mucous secretions of the intestinal tract to normal. It is a useful cleansing tonic for the liver. Its stimulant properties are well-known. It is also a valuable agent for arresting hemmorrhage of the uterus, bowels and lungs.

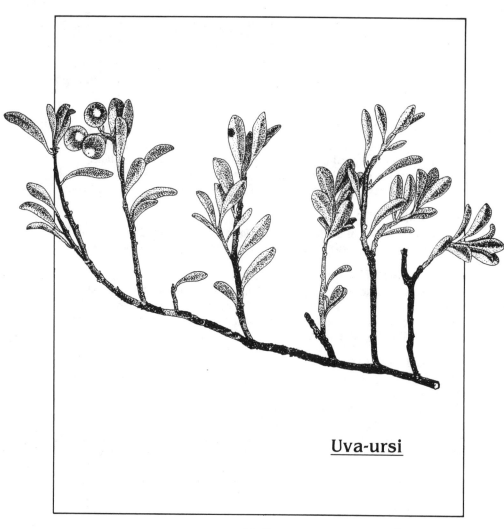

Uva-ursi

Uva-ursi or Kinnikinnick

Arctostaphylos uva-ursi
(Ericaceae)

Description:

Uva-ursi, also called kinnikinnick or bearberry, is a creeping to semi-erect shrub. Its long, flexible branches form mats by rooting; the twigs are covered with hairs. The leaves, about ¾ inch (2 cm.) long, are evergreen, leathery, oblong to obovate, light green on the bottom with no brown spots (see lowbush cranberry), and net-veined. The urn-shaped flowers vary from white to pink and the fruit is a mealy red berry, seedy and dry but edible.

Distribution:

Mats of uva-ursi are common in the boreal forest region of Alaska and across Canada in a wide variety of habitats. This plant is also found in northern Europe and Asia.

Constituents:

The chief medicinal principle of uva-ursi tea is a glycoside known as arbutin, found in many members of the heather family. (Turner and Czszawinski) Its other elements are methylarbutin, ursolic acid, tannic acid, gallic acid, some essential oil, and resin. (Pierson)

Medicinal uses:

The leaves are grown commercially for use as a diuretic and astringent, and the medicine, marketed under the name Uva Ursi or Bearberry, has long been used for relieving kidney and bladder problems

Certain constituents of the uva-ursi leaf combine with chemicals normally found in urine to form hydroquinone (P-dihydroxybenzene), which is bactericidal. Don't worry if the urine turns bright green. (Gibbons) Uva-ursi helps to reduce accumulations of uric acid and to relieve the pain of bladder stones and gravel; use it to alleviate cystitis.

The tea is known as *Kutai,* or "Caucasian tea," in Russia. It is popular in many parts of the world as a tonic or health tea, good for stomach and urinary disorders such as bedwetting. Use 1 teaspoon of dried leaves per cup of boiling water and steep for five minutes. Prospectors and trappers recommend soaking the leaves in whiskey first and then using them in the normal way to make tea. (Turner and Czszawinski) The tea or tincture can be used for bronchitis, according to Lust.

The term *kinnikinnick* is said to be derived from an Algonquin Indian word meaning "smoking mixture." The name has come to be used for *Arctostaphylos uva-ursi,* whose leaves are often used in smoking mixtures. (Turner and Czszawinski) The Chippewas smoked these leaves as a headache cure.

Lime villagers say uva-ursi berries are good as a laxative. (Kari)

Caution: Excessive, prolonged use should be avoided; in any usage both an aromatic and a demulcent should be added to the leaves, since their action when alone is too irritating.

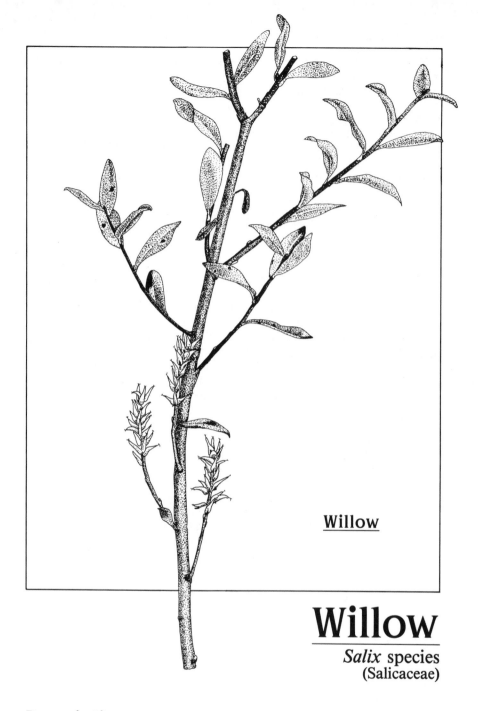

Willow

Willow
Salix species
(Salicaceae)

Description:

The willow catkin, or pussy willow, is a familiar feature of the spring land-scape in shrubby areas. Some willows flower before the leaves open out, while in other species the leaves appear first. The willows vary from prostrate or creeping dwarf shrubs to erect bushes or small trees, usually with many stems. The short-stalked leaves are long and narrow with smooth or finely toothed edges.

Distribution:

The several dozen species of *Salix* that occur in Alaska are widely distributed.

Constituents:

The active principle in willow bark is salicin, approximately 0.3 to 1%. Salicin is a glycoside hydrolyzed to D-glucose and saligenin that is a precursor to salicylic acid. The bark also contains tannins.

Medicinal uses:

The history of aspirin begins with the willow, which was used in Europe for fevers, debilities of the digestive system, bad scorbutic tumors, and dysentery. (Simmonite-Culpeper) The European use of willow bark infusions dates back to ancient Greece, where it was used to treat pain more than 2400 years ago.

In the mid-eighteenth century, Reverend Edmund Stone, in a letter to the president of the Royal Society, gave "an account of the success of the bark of the willow in the cure of agues [fever]." It was in 1837 that Leroux discovered or isolated salicin as the active ingredient in willow bark. In 1860 the synthetic manufacture of this acid from phenol was accomplished by Kolge and Lautemann. (Goodman and Gilman) Before 1900, sodium salicylate, pheylsalicylate, and acetylsalicylic acid (aspirin) were synthesized and introduced into medicine, soon displacing the natural compounds.

In Alaska, Oswalt reports the use of *S. arbusculoides* as a chewing stick for sores in the mouth. I have tried it, and this willow is certainly one of the most likely species to want to chew on — the shiny red bark of the slender twigs and branches is very smooth. *S. pulchra* also has been used for a sore mouth, but the leaves are chewed. (Lewis) The beautiful willow *S. pulchra* is the species most highly regarded as spring food; the young green leaves are good either raw or cooked. They seem to be preferred by moose too, because often when I go to one of the *S. pulchra* bushes to harvest spring buds it is eaten down to short sticks with all the buds gone and moose-type teeth marks on the cut ends.

Lewis and Elvin-Lewis have much to say about willows. Toothpicks made from various willows are commonly used. Comanches used a decoction of willow for sore eyes. The Houmas and Alabamas used a decoction of the root and bark of *S. nigra* for internal consumption and for baths to reduce fevers. The Chickasaws used the roots of *S. lucida* for headaches and the Montagnais steeped its leaves and drank the liquid to relieve their headaches. They also made a mash of the bark, which was strapped to the forehead to relieve pain.

Alaskan Native Martha Jack's informant suggests a cough syrup made from willow buds, and Hall states "When a bee stings you, get some willow leaves and chew them up for a few minutes. Put the chewed leaves over the sting. The leaves help keep the sting from swelling." She also says to bathe skin infections in willow leaf tea.

Caution: With willow, there is a possibility of skin rashes. However, salicylate poisoning has certainly occurred with synthetic aspirin, but has never been reported from natural willow bark.

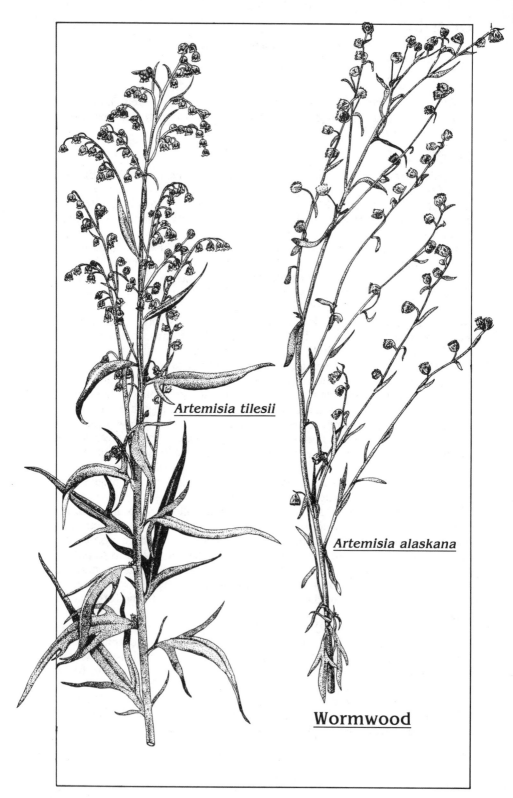

Artemisia tilesii

Artemisia alaskana

Wormwood

76

Wormwood

Artemisia species
(Compositae)

Description:

The genus *Artemisia* includes the wormwoods, absinth, sagebrush, mugwort, and tarragon — all pungently aromatic and bitter herbs and shrubs. They are perennials with small but numerous heads of the the composite type, borne in clusters more or less branched, in a spike, raceme, or panicle.

Artemisia alaskana flowers are yellow and the leaves along the stem (see illustration below) are blunt-tipped and twice ternate. The flowering stem rises from prostrate branches grown the previous year. Hairs cover the white-silvery stems and leaves.

Artemisia tilesii stems are more erect, rising directly from the woody base of the plant to a height of 2 to 3 feet (.7 to 1 meter). The leaves are not silvery white; they're green and hair-covered only on the lower surface. The leaves of *A. tilesii* usually divide into narrow lobes having slender, sharp tips. The flowers are yellowish brown. Both *A. alaskana* and *A. tilesii* have basal leaves that tend to be the largest of the plant's leaves. In *A. alaskana* the basal leaves can form rosettes.

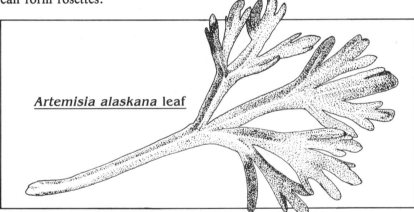

Artemisia alaskana leaf

Distribution:

Artemisia alaskana occurs mostly in the mountains in Alaska and adjacent Canada. *Artemisia tilesii* grows in sandy places in mountains and lowlands throughout most of Alaska and into Canada and Siberia.

Constituents:

The common wormwood or absinth, *A. absinthium,* contains a dark green or blue volatile oil with a strong odor and bitter taste. The oil contains absinthol or tenacetone, thujyl alcohol, cadinene, phellandrene, and pinene. The herb also contains the bitter glucoside absinthin, absinthic acid, together with tannin, resin starch, nitrate of potash and other salts. (Grieve)

Werner Herz of Florida State University has reported the isolation of three sesquiterpene lactones from *A. tilesii*. The plant also contains artilesin, which is bactericidal.

Medicinal uses:

The volatile oil of common wormwood or absinth, *A. absinthium,* is a central nervous system depressant causing trembling then stupor, followed by convulsions. (Spoerke) It has been abused and is habit-forming. However, Christopher considers it a valuable tonic that stimulates appetite and promotes digestion. There seems to be widespread agreement that the leaves possess antiseptic properties and are very resistant to putrefaction. The name may be a reflection of the idea that wormwood leaves and flowers expel worms.

Alaskans use native wormwoods both externally and internally. Both Oswalt and Lantis report the use of *A. alaskana* on the rocks in steam baths. I do this myself and enjoy the aromatherapy both for sinuses and skin. Priscilla Kari says the Tanainas still use *A. tilesii* in the steam bath. I have tried it and find I prefer the stronger aroma of *A. alaskana,* but both species are pleasant to use.

The Tanainas soak *A. tilesii* leaves in water and rub them on the bodies of pregnant women or put them on the stomach as a poultice. They also make medicine switches to help arthritis and other aches. Boiled or soaked in hot water, *A. tilesii* is made into a tea used as a wash for skin rash, cuts, blood poisoning, sore eyes, or any kind of infection. Use boiled or soaked leaves wrapped in cloth as a hot pack for toothache, earache, and snow-blindness. For athlete's foot, the Outer Cook Inlet people wear fresh leaves inside their socks. (Kari)

Artemisia tilesii is one of the medicinal herbs used by Della Keats (a respected healer) in Kotzebue, and by the people of the northwestern region of Alaska. It is highly regarded as a tonic tea if you don't drink too much at a time. Dried leaves are powdered to use externally in a salve for burns or infections. *Artemisia tilesii* was used by western Eskimos as an antitumor agent in Unalakleet and as a fever and infection inhibitor in Aniak, according to Smith.

Hall's *Traditional Medical Practices* of the Interior people includes the following uses for wormwood: "Use moistened, dried leaves as poultice on infected sores or cuts. Use just enough to cover sore. Put it on sore and wrap with bandage. Use as directed for arthritis also. Brew just a pinch of leaves and drink ½ cup once a day. Use for diaper rash also."

Brøndegaard surveys the use of *Artemisia* throughout the world in gynecological folk medicine. It is one of the medical plants used in Scandinavia, Germany, France, Switzerland, England, Bosnia, Russia, China, Tibet, India, Bali, Bolivia, Argentina, and the United States. It has been widely used to help women regulate menstruation and recover from childbirth.

The numerous uses of *Artemisia* for women may be related to the fact the plant is named for Artemis, the goddess who represents the variable energies of women. (Monaghan)

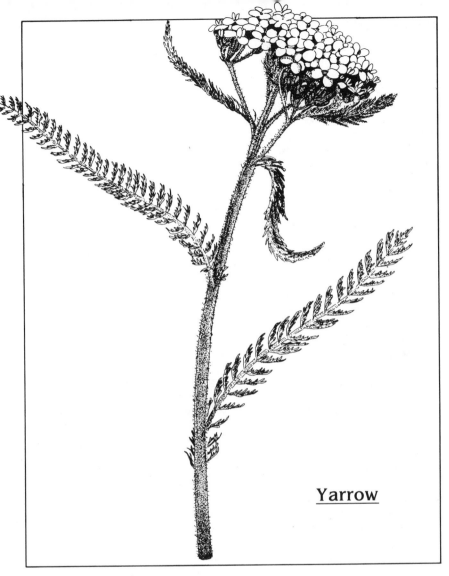

Yarrow

Yarrow

Achillea millefolium
(Compositae)

Description:

Yarrow's erect stalk, covered with appressed hairs, rises ½ to 4 feet (45 to 120 cm.) tall. The narrowly oblong leaves, with finely dissected segments (hence one of the common names, thousand leaf) along the stem, form a basal rosette spread along the ground. The white or grayish-white flowers, numerous and small, form on one or more branching heads in the form of a corymb or cyme. The root is slender and branched.

79

Distribution:

Found in Europe and the Americas, forty species belong in the yarrow genus.

Constituents:

Achillea millefolium contains a volatile oil with cineol, a tannin, achilleine, achilletin, ivain, aconitic acid, stachydrin, choline, and glycocoll betaine. (Merck) B-sitosterol and achillin, a lactone, were isolated from this species by Tewari Srivastava and Bajpai. Three new flavones were isolated from this species by Falk et al. in 1975, and there is a bitter caledivain.

Khafagy et al. have characterized santolin, the bitter principle of *A. santolina* growing in Egypt. The pharmacological properties of santolin are similar to digitalis.

Medicinal uses:

In many systems of medicine, including Indian (Ayurvedic), European, Egyptian, and Amerindian, yarrow has been used as a tonic and stimulant to induce perspiration and reduce fever. It has also been considered a diuretic, astringent, emmenagogue, and vulnary. (Christopher) Its pungent aroma has a quality I describe as "medicinal," and my students agree for the most part, although agreement on odors is even more difficult than agreement on tastes.

In Alaska, Natives boil yarrow and put it, while hot, on swollen infections. (de Laguna) They make a hot pack out of the cooked or raw wet leaves, then put the pack on an ache, pain or sore. Priscilla Kari reports that the Tanainas dry the leaves and pound them into a powder, then put the powder on a sore, cut, burn or blister. (This use of the powder for burns or cuts was described to me by an herbalist who lives in Ambler, Anore Jones.) Tanainas also boil or soak the above-ground yarrow plant in hot water. Then they give this tea to a new mother; it cleans one out like chamomile. In Kenai, yarrow is a medicine for stuffed-up sinuses: boil the plant in water and inhale the steam.

Yarrow is the herb tea of my choice for the common cold, especially if the hot tea is sipped slowly while inhaling the vapors of the tea and a decongestant such as balm of Gilead. The victim should be comfortably wrapped in a blanket with the feet in a hot footbath of strong ginger root tea. (Viki, my friend and neighbor, makes a refreshing tea using mint from my garden with yarrow.) Other herbalists have emphasized the use of yarrow tea for colds.

For piles a yarrow fomentation gives relief, no doubt due to its astringent properties. A douche has been used for leucorrhea and other vaginal problems since yarrow is drying and binding. Even today the Aleuts pluck the leaves, roll them between their palms, and place them over open cuts as a coagulant. (Smith) Leaves are also crushed and stuffed into the nostrils for nosebleeds. However, I find myself sneezing a lot when I am crushing dried yarrow plants.

In regions adjacent to Alaska, Gunther has much to say about *A. millefolium:*

> Its aromatic properties were recognized by the Swinomish in its use as a bath for invalids, and the Quileute boiled the leaves in the room where an infant was sick to make the air smell pleasantly. The Cowlitz soak the leaves for a hair wash.
>
> For a stronger use Makah women eat the leaves raw to produce sweating at childbirth, boil them and drink the tea to purify the

blood, and drink a stronger solution to heal the uterus after birth. The Klallam use a similar tea during childbirth and for colds as well, mixing it for the latter with wild cherry bark. The Quinault boil the roots for tuberculosis and also use the tea as an eyewash. The Cowlitz and Squaxin believe the same tea is effective for stomach trouble. The Chehalis boil the leaves and drink the tea to stop the passage of blood with diarrhea. Before the coming of the whites they were subject to this illness from eating too much raw meat, according to one informant. The Skagit and Snohomish also use this diarrhea remedy.

The plant is used as a general tonic by the Quinault, by boiling the roots. The Lummi boil the flowers and drink the tea to relieve body aches, and one informant feels she did not get mumps from her children because of this use. This drink produces sweating, as does the Makah preparation used at childbirth.

Yarrow is also used as a poultice, the Klallam chewing the leaves and putting them on sores. The Squaxin smash the flower to use the same way. The Quileute lay the boiled leaves on rheumatic limbs and reduce fever with them . . .

In other parts of the world, Highlanders of Scotland still make a yarrow ointment to apply to wounds, piles, and the skin of sheep. Simmonite-Culpeper inform us that yarrow, also called soldier's woundwort and carpenter's weed, is an herb of Venus, famous for its wound-healing properties. When Linnaeus gave the genus its name, he was obviously knowledgeable about its mythological connection (this plant was presumably used by Achilles to heal the wounds of his soldiers in battle). I have often quipped that it doesn't seem to work well to heal the heel.

A Barefoot Doctor's Manual of China includes *A. sibirica,* which acts as a carminative and stomach tonic, clears meridian passages, and reduces inflammation. *Achillea santolina* was used by the Bedouins of the Egyptian desert. They say they use it in a steam bath to relieve rheumatic pains. It is one of the aromatic plants I often use on the rocks in my sauna.

The effect of yarrow on body temperature was investigated in India by Falk et al., who administered achillin to rabbits and subsequently measured a fall in rectal temperature. The mechanism is unknown, but the alkaloid reduces clotting time in rabbits. (Spoerke)

Therapeutic Uses of Alaskan Plants

This list of the therapeutic uses of Alaskan medicinal plants is simply a cross-reference to information in the text of the book. The reader should refer to the main entry for discussions of cautions or contraindications and for more information about the method of gathering and preparing the material. The information is for educational purposes only, and not intended as medical self-care advice; if you do use the medicines on yourself you do so at your own risk.

NOTE: The plant part used is the leaf, usually as tea, unless otherwise stated. The entries are in alphabetical order; darker type signifies the more important plant for the particular use.

Alterative — alder, chickweed, dandelion root and leaves, iris, Labrador tea, plantain, sourdock root

Aphrodisiac — angelica

Anesthetic — wormwood, yarrow

Anodyne — chamomile

Antibiotic — sphagnum

Anthelmintic — chamomile, iris, shield fern rhizome, wormwood

Antiscorbutic — chickweed, juniper berries, Labrador tea, poplar bark, rose hips and leaves, shepherd's purse, spruce needles, strawberry berries and leaves

Antiseptic — poplar (balm of Gilead), sphagnum, wormwood

Antispasmodic — chamomile, highbush cranberry bark

Anti-tumor — alder

Appetizer — **sourdock root** as a bitter tonic

Astringent — alder, birch, **bistort root,** cleavers, fireweed root, horsetail, plantain, **raspberry,** shepherd's purse, shield fern rhizome, sourdock root, strawberry root, uva-ursi, willow, yarrow

Baby bath — **rose leaf tea**

Bactericide — sphagnum, uva-ursi, wormwood

Bath house enhancer — chamomile, devil's club bark, **wormwood leaves, yarrow leaves**

Blood purifier — angelica, sourdock root, spruce needle tea

Bitter aromatic — chamomile

Bladder tonic — plantain, uva-ursi

Bowel tonic — cleavers

Cancer treatment — sourdock root

Carminative — angelica, cow parsnip roots, fireweed, yarrow

Cathartic — iris root, Labrador tea (large dose), poplar bark

Central nervous system depressant — wormwood

Cholagogue — dandelion

Circulatory system alterative — alder, plantain

Comforting tea — birch, fireweed, rose

Counterirritant — poplar

Cure-all — chamomile

Demulcent — chickweed, coltsfoot, poplar bark, shepherd's purse

Diabetic nutrition — devil's club root bark

Diaphoretic — angelica, chamomile, coltsfoot

Disinfectant — sphagnum

Diuretic — angelica, birch, **cleavers,** dandelion, fireweed, **horsetail,** iris, juniper berries, Labrador tea, poplar bark, strawberry roots, uva-ursi, yarrow

Emetic — alder, chamomile (large warm doses), devil's club stem bark, fireweed root, iris root, rose bark decoction (strong)

Emmenagogue — angelica, chamomile, yarrow

Emollient — coltsfoot

Expectorant — angelica, coltsfoot, poplar buds, poplar bark

Eye drops — alder bark infusion, chamomile tea, chickweed, currant stem bark tea, shield fern rhizome tea

Flu treatment — currant stem bark tea

Fomentation for chest — cleavers, coltsfoot leaf decoction, **poplar (balm of Gilead)**

Fumigation — juniper

Gargle for sore throat — alder, cow parsnip (whole plant), raspberry leaf tea, juniper berry

Glandular system alterative — plantain

Hepatic — dandelion, iris

Incense — juniper

Kidney tonic — cleavers, chamomile, crowberry juice, plantain, shield fern rhizome infusion, spruce pitch tea, uva-ursi

Laxative — birch (combined with alkali), chamomile, cleavers, dandelion root, iris root infusion, Labrador tea (strong), larch bark, sourdock root, uva-ursi

Leucorrhea treatment — plantain

Liniment — yarrow

Lymph system tonic — plantain

Menstrual pads — **sphagnum**

Mouthwash — angelica, birch leaves and bark, currant leaves

Nutritive — chickweed, cleavers, currant berries, highbush cranberry berries, poplar buds, rose hips, shepherd's purse, spruce needles

Peristaltic — poplar buds

Plasters — spruce tea

Poultice — angelica leaves, chickweed, cow parsnip (whole plant), currant root, horsetail, juniper

Poultice for aches, pains, sores, infections — angelica root, cleavers

Poultice for burns, wounds, bruises — birch bark

Poultice for chest congestion — coltsfoot

Poultice for pains and swellings — chamomile tea

Preventive tonic — **angelica root**

Purgative — devil's club stem bark, iris

Refrigerant — chickweed, cleavers

Resolvent — iris, poplar bark

Rubefacient — cow parsnip leaves and stem

Salve — chickweed

Sedative — chamomile, yarrow

Sialagogue — iris

Smoking mixture — coltsfoot, juniper, uva-ursi

Steam bath fragrance — devil's club, wormwood, yarrow

Stimulant — angelica, chamomile, cow parsnip root, iris, Labrador tea, poplar buds, shepherd's purse, wormwood, yarrow

Stimulant to digestive system — wormwood

Stimulant to genito-urinary tract — juniper

Stomachic — angelica, dandelion roots and leaves, juniper berries, poplar bark, wormwood, yarrow

Styptic — bistort root, chamomile, dandelion, juniper, poplar bark

Sweat producer — yarrow

Tonic — alder, **angelica root**, birch sap, bistort (tones alimentary tract), cleavers, coltsfoot, cow parsnip (whole plant), cranberry leaf, dandelion, juniper, Labrador tea, poplar bark, raspberry (especially good for pregnant women), shield fern rhizome, sourdock root, spring willow root tea, wormwood, yarrow

Vasodilator — juniper berries

84

Vulnerary — poplar (balm of Gilead), shield fern rhizome, spruce gum, yarrow

Wound dressing — dried inner bark of devil's club mixed with cedar or spruce pitch, spruce gum

Conditions That Can Be Treated with Alaskan Plants

Let me emphasize once again that this book is not intended to be a medical self-care manual. Some of the diseases in this list are so serious a physician should be consulted as promptly as possible. Some of the alleged plant uses may be apocryphal or unfounded. Others, however, may be valid and useful to you. I would advise you to proceed with intelligence, good common sense, and caution — at your own risk, of course.

NOTE: The plant treatments printed in darker type are thought to be the most effective for the particular condition.

Abcesses — **chickweed poultice**

Aches — cleaver or coltsfoot hot pack, **wormwood poultice**

Ague — angelica

Alcoholism — angelica

Arthritis — cow parsnip root, juniper (branch, stem, berries), **willow,** wormwood, yarrow bath

Asthma — coltsfoot, shield fern rhizome

Athlete's foot — wormwood (fresh leaves inside socks)

Back trouble — juniper, hot spruce pitch plaster

Bedwetting — uva-ursi

Beestings — plantain, willow leaf poultice

Bites — plantain

Bladder stones — horsetail, uva-ursi

Bladder trouble — tonic of plantain or uva-ursi

Bleeding — bistort, cleavers, larch bark, plantain, shepherd's purse

Blisters — yarrow

Blood poisoning — plantain, sourdock root (for heavy metals), spruce pitch poultice, wormwood tea

Bloody stools — horsetail

Boils — **chickweed,** devil's club root bark poultice, mashed green plantain leaves, sweet gale tea

Bowel trouble — **cleavers** (mild tonic)

Broken bones — **sphagnum** (for splint lining), dandelion greens (nutritional source of calcium)

Bronchitis — chamomile, **coltsfoot,** poplar, uva-ursi

Bruises — chamomile, poplar root poultice

Burns — chickweed, coltsfoot leaf decoction, devil's club ashes, yarrow

Cancer — sourdock root

Cankers — horsetail ashes

Chest trouble — coltsfoot, juniper, poplar (balm of Gilead)

Circulatory trouble — alder

Colds — **chamomile,** cow parsnip root, currant stem bark tea, devil's club root bark tea, highbush cranberry bark decoction, horsetail, juniper, Labrador tea, poplar, **rose hips,** rose twig decoction, spruce pitch tea, **yarrow**

Colic — angelica, chamomile

Congestion (to loosen) — bistort root, chamomile, coltsfoot

Conjunctivitis — horsetail tea

Constipation — cleavers, iris root infusion

Convulsions — highbush cranberry bark

Corns — chamomile

Cough (spasmodic) — chamomile, **coltsfoot,** juniper

Cramps — **highbush cranberry bark**

Cutaneous diseases — larch bark tea

Cuts — powdered bistort root, mashed and cooked sourdock root salve, raspberry stem, powdered wormwood leaves, soapberry stem tea, **spruce pitch,** wormwood tea, yarrow

Cystitis — **uva-ursi**

Debility — angelica

Delerium tremens — chamomile

Diabetes — angelica, cleavers, devil's club root bark

Diaper rash — **sphagnum,** wormwood

Diarrhea — alder, cleavers, coltsfoot (leaf or flower infusion), crowberry leaf and stem tea, fireweed root, horsetail, larch leaf tea, plantain, raspberry root tea, shepherd's purse, strawberry roots and leaves boiled in wine and water, sweet gale bark, **willow bark,** willow root tea enema, yarrow

Dropsy — birch

Dyspnea — yarrow

Earache — wormwood leaf hot pack, yarrow compress

Eczema — juniper

Edema — angelica, birch, **cleavers,** dandelion, fireweed, **horsetail,** iris, juniper berries, Labrador tea, poplar bark, strawberry roots, uva-ursi, yarrow

Erysipelas — coltsfoot leaf decoction

Eyes, sore — chickweed, coltsfoot, crowberry juice or stem bark tea, crowberry root decoction, currant stem bark tea, horsetail, rose petal juice, sphagnum (red kind), willow bark, wormwood, yarrow hot compresses

Famine — any berry except baneberry, horsetail root, inner bark of most trees

Fever — bistort root, chamomile, currant juice, devil's club root bark tea, poplar, rose twig decoction, shepherd's purse, sourdock, willow bark, yarrow

Fever (intermittent) — birch

Flatulence — angelica root infusion, juniper berry tea, yarrow

Flu — currant stem bark tea, horsetail, poplar

Foot trouble — alder

Freckles (to remove) — cleavers

Frostbite — poplar bud

Fungus infections — poplar bark

General poor health — chamomile

Glands, swollen — devil's club root bark tea

Gout — birch

Gut trouble — tea of shield fern rhizome or birch leaves

Hair problems — tonic of yarrow

Hangover — sourdock root decoction

Headache — lowbush cranberry hot pack, plantain crushed leaf poultice, poplar leaf compress, spruce pitch and snow poultice, uva-ursi

Headache, bilious — chamomile

Heartburn — angelica

Heart trouble — sourdock as a cordial

Hematuria — horsetail

Hemorrhoids — larch (inner bark tea), yarrow

Hemorrhages — bistort, shepherd's purse, yarrow

Haemoptysis — larch leaf tea

Hives — sourdock leaves, spruce needle wash

Hypertension — horsetail

Hysteria — chamomile

Indigestion — alder, angelica, chamomile

Infection — angelica root poultice, chickweed poultice, devil's club root bark poultice, hot sphagnum poultice, spruce pitch and grease, wormwood, hot yarrow poultice

Inflammation — chickweed, coltsfoot, shepherd's purse, yarrow

Insect stings — bistort root decoction, coltsfoot leaf decoction

Itching — sourdock juice, uva-ursi

Jaundice — sourdock root

Joint pains — wormwood, yarrow

Kidney stones — birch, cranberry, currant, horsetail, poplar bark, uva-ursi

Kidney trouble (tonic for) — cleavers, chamomile, crowberry juice, plantain, shield fern rhizome infusion, spruce pitch tea, uva-ursi

Leg ulcers — coltsfoot leaf decoction

Leucorrhea — strong bistort root decoction, plantain, yarrow

Liver, torpid — chamomile, dandelion root, larch bark tea, poplar bark

Mange — sourdock salve

Melancholy — yarrow

Menorrhagia — larch

Menstrual flow, difficult — currant leaves

Menstrual flow, to decrease — bistort root decoction douche, horsetail, larch bark tea, plantain

Menstrual flow, to increase — angelica root infusion tea

Menstrual flow, irregular — angelica, horsetail decoction with willow leaves, wormwood

Menstrual flow, painful — birch "cone" tea, chamomile, **highbush cranberry bark,** larch bark tea

Mosquitos (repellent) — elder, plantain, yarrow

Mouth problems (mouthwash) — angelica, birch leaves and bark, currant leaves

Mouth sores — cow parsnip root, willow bark

Muscle cramps — highbush cranberry bark

Nausea — raspberry leaf tea

Nervousness — chamomile, cow parsnip (eaten)

Nosebleed — shepherd's purse tea, yarrow powder

Obesity — cleavers, horsetail

Pain — poplar, sourdock root, willow bark, wormwood, yarrow root

Phlebitis — coltsfoot leaf decoction

Piles — larch bark tea

Pimples — sweet gale tea (used as wash)

Postparturition — chamomile, **raspberry leaf,** yarrow

Pregnancy — coltsfoot, **raspberry leaf tea** (for nausea), wormwood leaves (for massage)

Psoriasis — juniper

Rash — poplar, spruce needle tea wash

Restlessness — fireweed tea

Rheumatism — birch, chamomile, coltsfoot hot pack, coltsfoot root decoction, juniper, Labrador tea, larch bark, yarrow (on steam bath rocks)

Ringworm — plantain, sourdock

Runny nose — raspberry root tea

Scratches — spruce gum

Seasickness — angelica root fumes

Sinus trouble — inhale fumes from roasting birch cones or boiling yarrow

Skin, itching — plantain

Skin disorders — larch bark tea

Skin eruptions — birch leaf tea, chamomile, plantain, sweet gale tea

Skin infections — bathe in willow leaf tea

Sore mouth or gums — bistort root infusion, raspberry leaf tea

Sore throat — alder bark, coltsfoot, cow parsnip, highbush cranberry leaf decoction, larch bark tea, poplar bark, rose root decoction with sugar, willow bark

Sore throat (gargle for) — alder, cow parsnip (whole plant, raspberry leaf tea

Sores — bistort root infusion, chamomile tea, powdered currant stem ashes, devil's club root bark poultice, hot juniper berries, sourdock salve, ashes of burned willow twigs, yarrow powder

Spleen trouble — chamomile

Sprains — chamomile, poplar root poultice

Steam bath — devil's club, wormwood, yarrow

Stings — plantain, willow leaf poultice

Stomach, gas in — alder

Stomach, upset — lowbush cranberries, strawberries

Stomachache — currants (eat the berries), devil's club root bark tea, fireweed tea

Stomach trouble — rose twig decoction, sourdock root tea, uva-ursi, yarrow

Stomach ulcers — coltsfoot

Stomach weakness — chamomile, highbush cranberry bark decoction

Sunburn — poplar (balm of Gilead)

Swelling (to reduce) — angelica root, bistort root, coltsfoot decoction, soapberry stem tea, sourdock root, wormwood poultice

Toothache — angelica root, bistort, chamomile, hot juniper poultice, wormwood, yarrow leaves (chew them)

Tumors — alder stem bark, plantain poultice, shepherd's purse

Ulcers — chamomile, dandelion, juniper, plantain, sourdock root

Urinary problems — angelica, sitz bath of spruce needle tea, uva-ursi

Urine, scanty — birch bark and leaves, uva-ursi

Vermin — horsetail shampoo

Vomiting — raspberry leaf tea

Warts — spruce pitch

Weak blood — rose twig decoction

Weight (to lose) — cleavers

Worms (to expel) — angelica, chamomile, juniper

Wounds — powdered bistort root (applied externally), chickweed, horsetail (for internal wounds of the bowels), hot juniper berries, shepherd's purse (externally as a poultice), yarrow leaves (as an ointment)

Glossary

Alterative	An agent that gradually restores healthy bodily functions.
Alternative	Same as alterative.
Amenorrhea	Absence of menstruation.
Anther	The pollen-production portion of the stamen.
Anodyne	An agent that allays or kills pain.
Antiscorbutic	A remedy for scurvy.
Appressed	Lying flat or close against something.
Astringent	An agent producing contraction of tissue or arrest of discharge.
Aromatherapy	Using odors to heal.
Avatar	The embodiment of a deity.
Bradycardia	Abnormally slow heart rate.
Bronchus	One of the main branches of the trachea.
Calyx	The sepals, collectively.
Carminative	A medicine that promotes expulsion of intestinal gas.
Catarrh	Inflammation of a mucous membrane.
Cathartic	An agent producing watery evacuations; a purgative.
Corolla	The petals, collectively.
Cordate	Heart-shaped base.

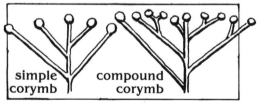

Corymb	A flat-topped, open inflorescense.

Cuneate	Wedge-shaped, triangular.
Cyme	A flower cluster in which the central or terminal flower blooms first.
Decoction	An infusion made in boiling water.
Demulcent	A mucilaginous substance allaying irritation.
Depressant	Having the effect of decreasing vital activity.

Depurant	A cleansing, purifying agent or drug.
Diaphoretic	An agent producing perspiration.
Dysmenorrhea	Pain during menstruation.
Emetic	Agent used to bring on vomiting.
Emmenagogue	Agent stimulating menstrual flow.
Emollient	Agent that softens tissues.
Epidermis	Outer layer of cells.
Essential oil	Natural oil as opposed to synthetic oil.
Expectorant	Agent that promotes the secretion of bronchial mucus.
Filament	The stalk of the anther portion of the stamen.
Fomentation	The application of warm liquids to the body.
Glandular	Bearing glands (secreting agents).
Glandular hairs	Hairs bearing a swelling at the tip.

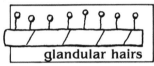

Globose	Shaped like a globe or a sphere.
Gravel	Sandlike deposit in the urine.
Hallucinogen	Having the psychoactive effects of producing imaginary perceptions.
Inflorescence	A flower cluster.
Infusion	Liquid prepared by steeping or soaking a drug in water.
Involucre	A whorl of leaves beneath a flower or inflorescence.
Lax	Loose.
Lanceolate	Lance-shaped, several times longer than wide; broadest toward the base.

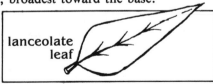

Leucorrhea	Inflammation of the vaginal or uterine mucosa, usually characterized by a whitish or yellowish vaginal discharge.
Lenticel	A group of loose corky cells formed beneath the epidermis of woody plants, rupturing the epidermis and admitting gases to and from the inner tissues.
Mucous membrane	The thin lining of those cavities and canals communicating with the air.

Mucus	The viscid secretion of mucous membrane.
Oblanceolate	Opposite of lanceolate — broadest toward the tip.

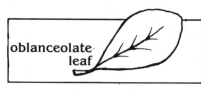

Oblong	Two to four times longer than wide; the sides are parallel.

Obovate	Inversely egg-shaped; attached at the narrow end.

Ovate	Egg-shaped.

Panicle	A compound raceme type of inflorescence.

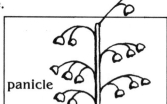

Pectoral	A remedy for chest diseases.
Petals	The corolla or inner floral envelope; variously colored.
Petiole	The stalk of a leaf.
Phlegm	A watery humor; mucus from the bronchi.
Pinna	One of the first divisions of a pinnately compound leaf.
Pinnate	Compound leaf with leaflets on two opposite sides of an elongated axis.

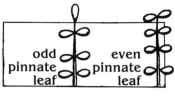

Pinnule	One of the second divisions of a bi-pinnately compound leaf.

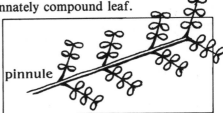

Pistil	The female organ of a flower.

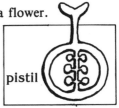

Pollen	The male spores.
Poultice	A soft mass, usually moist and heated, spread on a porous cloth and applied to an inflamed area.
Psychoactive	Having an effect on the mind.
Puberulent	With very short hairs.
Pubescent	Covered with hairs.

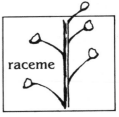

Raceme	An inflorescence with stalked flowers borne on a main axis.

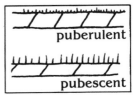

Rachis	The central axis of an inflorescence or compound leaf.
Rubefacient	An agent that reddens the skin.
Scrofula	A condition with tumors.
Scurvy	A deficiency disease due to low dietary intake of vitamin C; initial symptoms include loosening of teeth and damage to the glands.
Sepal	One of the parts of the outer floral envelope. Usually green.

Sepals	Parts of the calyx, usually green; the outer floral envelope.
Serrate	With sharp teeth directed forward.

Sessile	Without a stalk.
Spike	An inflorescence with the flowers on a straight axis.

Stamen	A pollen-bearing organ of a flower made up of anther and filament.

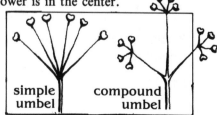

Stimulant	Having the effect of increasing vital activity.
Stipule	A modified leaf at the base of a bud.
Styptic	A medicine that causes contraction of the blood vessels and stops bleeding.
Subglobose	Almost shaped like a globe.
Ternate	Arranged in threes.
Tonic	An agent that produces normal tone or tension.
Umbel	A round or flat-topped inflorescence; the youngest flower is in the center.

Vulnerary	An agent useful in healing wounds.

References

Ager, Tom and Ager, Lynn Price. 1980. An Ethnobotany of Nelson Island. *Arctic Anthropology* 17(1):27-48.

Airola, Paavo. 1974. *How to Get Well. Dr. Airola's Handbook of Natural Healing.* Phoenix: Health Plus, Publ.

Anderson, J.P. 1939. Plants Used by the Eskimos of the Northern Bering Sea & Arctic Regions of Alaska. *Am. J. Bot.* 26:714-716.

Angier, Bradford. 1978. *Field Guide to Medicinal Wild Plants.* Harrisburg, PA: Stackpole Books.

Bank, Theodore. 1950. Botanical and Ethnobotanical Studies in the Aleutian Islands. *Pap. Mich. Acad. Sci. Arts, Letters* 36:13-30.

A Barefoot Doctor's Manual: The American Translation of the Official Chinese Paramedical Manual. 1977. Philadelphia: Running Press.

Barnes, C.S.; Price, J.R.; and Hughes, R.L. 1975. An Examination of Some Reputed Antifertility Plants. Lloydia 38(2) Mar-Apr:135-140.

Benoit, P.S.; Fong, H.H.; Svoboda, G.H.; and Farmsworth, N.R. 1976. Biological & Phytochemical Evaluation of Plants. XIV. *Antiinflammatory evaluation of 163 species of plants. Lloydia* 39(2-3) Mar-Jun:160-171.

Brondegaard, V.J. 1965. Artemisia in Gynecological Folk Medicine. *Svensk Farmaceutisk Tidskrift,* vol. 22, no. 23.

Carroll, Ginger. 1972. Traditional Medical Cures Along the Yukon. *Alaska Medicine.* April:50-53.

Christopher, John R. 1976. *School of Natural Healing.* Provo, UT: Biworld.

Creekmore, Hubert. 1966. *Daffodils are Dangerous.* New York: Walker and Co.

Culpeper, Nicholas. *See* Simmonite

de Laguna, Frederica. 1972. *Under Mount Saint Elias: The History and Culture of the Yakutat Tlingit.* Washington, D.C.: Smithsonian Institution Press.

Densmore, Francis. 1944. Uses of Plants by the Chippewa Indians. *Ethnologisches Anzeiger* 44:280-370.

Egan, E.W. 1962. *Herb Identifier and Handbook.* New York: Sterling Publ. Co.

Elbert, Virginia F., and Elbert, George A. 1974. *Fun with Growing Herbs Indoors.* New York: Crown Publ.

Falk, A.J.; Smolenski, S.J.; Bauer, L.; and Bell, C.L. 1975. Isolation and Identification of Three New Flavones from Achillea Millefolium L. *Journal of Pharmaceutical Sciences* 64(11) Nov:1838-1842.

Felkova, M. and Jasicova, M. 1978. Substances Contained in Matricaria Chamonilla. *Ceskoslovenska Farmacie* 27(8):359-366.

Fielder, Mildred. 1975. *Plant Medicine and Folklore*. New York: Winchester Press.

Gibbons, Euell. 1962. *Stalking the Wild Asparagus*. New York: David McKay.

Gibbons, Euell. 1966. *Stalking the Healthful Herbs*. New York: David McKay.

Gilman, Alfred Goodman; Goodman, Louis S., and Gilman, Alfred. 1980. *The Pharmacological Basis of Therapeutics*. Sixth Ed. New York: Macmillan.

Grieve, Mrs. M. 1971. *A Modern Herbal*. 2 vols. New York: Dover.

Grigson, Geoffrey. 1959. *An Herbal of All Sorts*. London: The Aldine Press.

Gunther, Erna. 1975. *Ethnobotany of Western Washington*. Seattle: University of Washington Press.

Hall, Alan. 1976. *The Wild Food Trail Guide*. New York: Holt, Rinehart, and Winston.

Hall, Brenda. 1980. *Traditional Medical Practices*. Fairbanks, Alaska: Tanana Chiefs Conference.

Harris, B.C. 1972. *The Complete Herbal*. Barre, MA: Barre Publishing.

Hartwell, Jonathan L. 1971. Plants Used Against Cancer, a Survey. *Lloydia* 34:204-255.

Heller, Christine. 1974. *Wild, Edible, and Poisonous Plants of Alaska*. Fairbanks: University of Alaska Extension Service Publication No. 40.

Hocking, George M. 1955. *A Dictionary of Terms in Phamacognosy: Natural Medical, and Pharmaceutical*. Springfield, IL: Charles C. Thomas.

Horster, H. 1974. Variabilitat der Ole von Juniperus Communs. II. Die Zusammensetzung der Ole Reifer und Unreifer Fruchte. *Planta Medica* 25(1)Feb: 73-79.

Hsu, Hong-Yen. 1980. *How to Treat Yourself with Chinese Herbs*. Los Angeles: Oriental Healing Arts Institute.

Hulten, Eric. 1968. *Flora of Alaska and the Neighboring Territories*. Palo Alto: Stanford University Press.

Hu-nan. 1977. *A Barefoot Doctor's Manual. The American Translation of the Official Chinese Paramedical Manual*. Philadelphia: Running Press.

Hylton, William H. 1974. *The Rodale Herb Book*. Emmaus, PA: Rodale Press.

Jack, Martha. 1971. *Uses of Plants by Alaskan Natives*. Fairbanks: University of Alaska Museum. Unpublished, special topics.

Jackson, Mildred and Teague, Terri. 1975. *The Handbook of Alternatives to Chemical Medicine*. Oakland, CA: Lawton-Teague Publishers.

Jacob, Dorothy. 1965. *A Witch's Guide to Gardening*. New York: Taplinger Publ. Co.

Kadans, Joseph M. 1972. *Encyclopedia of Medicinal Herbs*. New York: Arco Publishing.

Kari, Priscilla Russell. 1977. *Dena'ina K'et'una. Tanaina Plantlore*. Anchorage: University of Alaska.

Khafagy, Saad. 1965. Preliminary Characterization of Santolin, Bitter Principal of Achillea Santolina L. Growing in Egypt. *Acta Pharmaceutica Suecica*. vol 2, no. 6.

Kingsbury, John M. 1964. *Poisonous Plants of the United States and Canada*. Eaglewood Cliffs, N.J.: Prentice-Hall.

Kirk, Donald R. 1975. *Wild Edible Plants of the Western United States*. Happy Camp, CA: Naturegraph Publ.

Kloss, Jethro. 1975. *Back to Eden*. Santa Barbara: Woodbridge Press.

Kreig, Margaret. 1964. *Green Medicine*. Chicago: Rand McNalley.

Kulvinskas, Viktoras. 1977. *Survival Into the 21st Century. Plantary Healer's Manual*. Wethersfield, CT: Omangod Press.

Kumar, R.; Banerjee, S.K.; and Handa, K.L. 1976. Coumarins of Heracleum Canescens and Heracleum Pinnatum. Sources for Dermalphotosensitizing Agents. *Planta Medica* 30(3) Nov:291-294.

Kuroda, K.; Akao, M.; Kanisawa, M.; and Miyaka, K. 1976. Inhibitory Effect of Capsella Bursa-Pastoris Extract on Growth of Ehrlich Solid Tumor in Mice. *Cancer Research* 36(6) June:1900-1903.

Kuroda, K., and Takagi, K. 1969. Studies on Capsella Bursa Pastoris. I. General Pharmacology of Ethanol Extract of the Herb. *Archives Internationales de Pharmacodynamie et de Therapie* 178(2) April:382-391.

Kuroda, K., and Takagi, K. 1969. Studies on Capsella Bursa Pastoris. II. Diuretic, *Anti-inflammatory and Anti-ulcer Action of Ethanol Extracts on the Herb. Archives Internationales de Pharmacodynamie et de Therapie* 178(2) April:392-399.

Kuroda, K., and Akao, M. 1975. Effect of Capsella Bursa Pastoris on Liver Catalase Activity in Rats Fed 3-Methyl 4 (Dimethylamino)-Azobenzene. *Gann* (Japan) V. 66 #4: 461-462.

Lantis, Margaret. 1959. Folk Medicine and Hygiene. Lower Kuskokwin and Nunivak-Nelson Island Areas. *Anthro. Paper* (University of Alaska, Fairbanks) 8:1-75.

Lehner, Ernst. 1960. *Folklore and Symbolism of Flowers, Plants, and Trees*. New York: Tudor.

LeStrange, Richard. 1977. *A History of Herbal Plants*. New York: Arco Publ. Co.

Levy, Juliette de Bairacli. 1976. *Herbal Handbook for Farm and Stable*. Emmaus, PA: Rodale Press, Inc.

Lewis, Walter H. 1977. *Medical Botany: Plants Affecting Man's Health*. New York: John Wiley & Sons.

Lucas, Richard. 1972. *Common and Uncommon Herbs for Healthful Living*. West Nyack, NY: Parker Publ. Co.

Lust, John. 1974. *The Herb Book*. New York: Bantam Books.

Marks, Geoffrey, and Beatty, W.K. 1971. *The Medical Garden*. New York: Scribner.

Majno, Guido. 1975. *The Healing Hand. Man and Wound in the Ancient World*. Cambridge, MA: Harvard University Press.

The Merck Index. An Encyclopedia of Chemicals and Drugs. 1968. Rahway, NJ: Merck and Co.

Millspaugh, Charles. 1974. *American Medicinal Plants*. New York: Dover.

Morita, N.; Shimizu, M.; Arisawa, M.; and Kitanaka, S. 1974. Studies on Medicinal Resources. XXXV. The components of Salix Plants (Salicaceae) in Japan. *Journal of the Pharmaceutical Society of Japan*. 94(7):875-877.

Moulton, Le Arta. 1979. *Herb Walk*. Provo, Utah: The Gluten Co.

Muir, Ada. 1959. *The Healing Herbs of the Zodiac*. St. Paul: Llewelyn Publ.

Nelson, Richard K. 1980. Athapaskan Subsistence Adaptions in Alaska. Alaska Native Culture and History. *Senri Ethnological Studies 4*, Osaka, Japan.

Niemann, G.J. 1974. Phenolics from Larix Needles. *Planta Medica*. 26(2) Sep:101-103.

Oswalt, W.J. 1957. A Western Eskimo Ethnobotany. *Anthro. Paper* (University of Alaska, Fairbanks) 6:17.

Padula, L.A.; Rondina, R.V.; and Coussio, J.D. 1976. Quantitative Determination of Essential Oil, Total Azulenes and Chamazulene in German Chamomile Cultivated in Argentina. *Planta Medica*. 30(3) Nov:273-280.

Physicians Desk Reference. 1983. 37th edition. Oradell, NJ: Medical Economics Co.

Pond, Barbara. 1974. *A Sampler of Wayside Herbs. Rediscovering Old Uses for Familiar Wild Plants*. Riverside, CT: Chatham Press.

Porteous, Alexander. 1968. *Forest Folklore, Mythology, and Romance*. London: Allen & Unwin.

Potter, Loren. 1972. Plant Ecology of the Walakpa Bay Area, Alaska. *Arctic* 25(2):115-130.

Prensky, Joyce. 1976. *Healing Yourself*. P.O. Box 752, Vashon, WA.

Rodahl, Kaare. 1952. Vitamin Content of Arctic Plants and Their Significance in Human Nutrition. Arctic Aeromedical Laboratory. *Trans. and Proc. Bot. Soc. Edinburgh* 36:267-277.

Salem, M. 1972. Effects of Light upon Quantity and Quality of Matricaria Chamomilla L. Oil. *Pharmazie* (East Germany) 27(9):608-611.

Schantz, M. Von and Kapetanidis, I. 1971. Qualitative and Quantitative Study of Volatile Oil from Myrica Gale L. (Myricaceae). *Pharmaceutica Acta Helvetiae* (Switzerland) 46(10):649-656.

Scultes, Richard Evans and Hofman, A. 1973. *The Botany and Chemistry of Hallucinogens.* Springfield, IL: Charles C. Thomas.

Sheth, K.; Bianchi, E.; Wiedhopf, R.; and Cole, J.R. 1973. Antitumor Agents from Alnus Oregona (Betulaceae). *Journal of Pharmaceutical Sciences* 62(1):139-140.

Simmonite, W.J. and Culpeper, N. 1957. *The Simmonite-Culpeper Herbal Remedies.* London: W. Foulsham & Co., Ltd.

Smith, Warren G. 1973. Arctic Pharmacognosia. *Arctic* 26(4):324-333.

Soden, Margaret. 1976. *First Aid for Poisonous Plants and Mushrooms.* Fairbanks: University of Alaska Extension Service.

Spoerke, David. 1980. *Herbal Medications.* Santa Barbara: Woodbridge.

Swain, Tony. 1972. *Plants in the Development of Modern Medicine.* Cambridge, MA: Harvard University Press.

Szczawinski, Adam F. and Turner, Nancy J. 1978. *Edible Garden Weeds of Canada.* Ottawa: National Museum of Canada.

Szczawinski, Adam F. and Turner, Nancy J. 1980. *Wild Green Vegetables of Canada.* Ottawa: National Museum of Canada.

Taskinen, J. and Nykanen, L. Chemical Composition of Angelica Root Oil. 1975. *Acta Chemica Scandinavica.* Series B. Organic Chemistry and Biochemistry. 29(7):757-764.

Tatum, Billy Joe. 1976. *Wild Foods Cookbook and Field Guide.* New York: Workman Publishing Co.

Tetenyi, Peter. 1970. *Infraspecific Chemical Taxa of Medical Plants.* Hungary. New York: Chemical Publ. Co.

Tewari, J.P.; Srivastava, M.C.; and Bajpai, J.L. 1974. Phytopharmacologic Studies of Achillea Millefolium Linn. *Indian Journal of Medical Sciences* 28(8):331-336.

Thompson, W.A.R. 1978. *Medicines from the Earth: A Guide to Healing Plants.* New York: McGraw-Hill.

Thiselton-Dyer, T.T. 1968. *The Folklore of Shakespeare.* New York: Dover.

Tobe, John H. 1975. *Proven Herbal Remedies.* New York: Pyramid Books.

Train, Percy; Henrichs, J.R.; and Archer, W.A. 1957. Medicinal Uses of Plants by Indian Tribes of Nevada. *Contr. Towards a Flora of Nevada.* No. 45. Agric. Res. Serv., USDA.

Traven, Beatrice. 1974. *The Complete Book of Natural Cosmetics.* New York: Simon and Schuster.

Turner, Nancy J. and Szczawinski, Adam F. 1978. *Wild Coffee and Tea Substitutes.* Ottawa: National Museum of Canada.

101

Twitchell, Paul. 1971. *Herbs: The Magic Healers*. San Diego: The Illuminated Way Press.

UNESCO COURIER. Medicine's Green Revolution. July, 1979.
Sai'd, H.M. Avicenna. "Hearts and Flowers."
Attisso, Michael A. "Medicinal Plants Make a Comeback."
Pelt, Jean Marie. "Medicine's Green Revolution."
Ekong, D.E.Q. "African Medicinal Plants Under a Microscope."
Khundanova, Lydia. "A Medical Thesaurus from the Roof of the World."
Wei Wen. "A New Medicine Born of Tradition."
Piattelli, Mario. "Neptune's Pharmacopoeia."
Crabbe, Pierre. "Mexican Plants and Human Fertility."
Gottlief, O.R. "Plant Prophylactics from Tropical Brazil."
Petkov, Vesselin. "Bulgaria's Folk Remedies Stand the Test of Time."

Viereck, Leslie A. and Little, Elbert L. 1972. *Alaska Trees and Shrubs*. U.S. Dept. of Agric. Handbook #410.

Viereck, Leslie A. and Little, Elbert L. 1974. *Guide to Alaskan Trees*. U.S. Dept. of Agric. Handbook #472.

Vohora, S.B.; Rizwan, M.; and Khan, J.A. 1973. Medicinal Uses of Common Indian Vegetables. *Planta Medica* 23(4):381-393.

Weinheim. 1966. On the Substances in the Rhizome of Polygonum Bistorta L. *Archiv der Pharmazie*. V.299 #7:640-646.

Index

104

U

Umbelliferae, 7-8, 23
Unalakleet, 78
Upper Cook Inlet people, 31, 35, 63
Ursolic acid, 73
Urtica species, 63
Uva-ursi, 72-73

V

Vaccinin, 47
Vaccinium vitis-idaea, 47
Valeric acid, 35, 44
Viburnin, 35
Viburnum edule, 35
Viburnum opulus, 35
Viopudial, 35
Vitamin A, 49
Vitamin C, 17, 49, 55, 69
Vitamin K, 57

W

Washington state, 27, 37, 55
Water hemlock, 8, 23
Waybread, 49
Weeds, 3, 29, 57
White pine, 27
Wild celery, 23
Willow, 74-75
World War I, 65
Wormwood, 3, 77-78

X

Xanthotoxin, 23

Y

Yarrow, 3, 79-81

About the Author

Eleanor G. Viereck earned a Ph.D. in biology from the University of Colorado. She wrote *Alaska's Wilderness Medicines* to serve as a text for the course she teaches in herb lore and holistic health at the Tanana Valley Community College, where she has taught since 1959. Her husband, a biologist for the U.S. Forest Service, wrote *The Trees and Shrubs of Alaska*. Their 20-acre tree farm and hand-built log home serve as a picnic stop and display arboretum for classes of Elderhostellers. She teaches and studies yoga.

Look to **Alaska Northwest Books**™ for other fascinating guides to the North Country, including:

Discovering Wild Plants: Alaska, Western Canada, The Northwest,
by Janice J. Schofield. Illustrated by Richard W. Tyler.
This beautiful book profiles 147 wild plants, providing definitive information on botanical identification, habitat, history, harvesting instructions, and recipes. Each plant is illustrated with color photos and precise line drawings.
Softbound, 368 pages, $26.95, ISBN 0-88240-369-9

Alaska-Yukon Wild Flowers Guide, edited by Helen A. White.
Wild flower lovers will appreciate this first Northland book to feature large color photographs of each plant and detailed line drawings for easy identification. A perfect travel companion for both residents and visitors to the Far North.
Softbound, 224 pages, $16.95, ISBN 0-88240-032-0

Alaska's Wild Plants: A Guide to Alaska's Edible Harvest,
by Janice J. Schofield.
For hikers, foragers, and plant lovers, *Alaska's Wild Plants* is an introduction to more than 70 common edible plants. This handy, pocket-size book shows you how to create delicious, nutritious meals from the land around you.
Softbound, 96 pages, $12.95, ISBN 0-88240-433-4

Alaska Wild Berry Guide and Cookbook
This is a book that delights the eye and the palate. The color photographs and drawings help you identify the berries; the recipes show you how to turn your harvest into scrumptious treats. A great gift for the whole family!
Softbound, 212 pages, $14.95, ISBN 0-88240-229-3

Ask for these books at your favorite bookstore, or contact
Alaska Northwest Books™.

Alaska Northwest Books™

An imprint of Graphic Arts Center Publishing Company
Catalog and Order Department
P.O. Box 10306
Portland, OR 97210
800-452-3032

What reviewers say about
Alaska's Wilderness Medicines:

"A handy field and kitchen reference . . . useful, backpack-size volume."
—Anchorage Times

"A handsome book that's also clear, thorough, well-organized, and fascinating to read . . . as interesting and potentially useful to casual readers as to specialists." —Fairbanks Daily News-Miner

"Packed with fascinating information about common plants."
—All-Alaska Weekly